SCIENCE
AND
SCIENTISTS

SCIENCE
AND
SCIENTISTS

Volume 3

Probability Theory - *Zinjanthropus*
Appendices
Indexes

from

The Editors of Salem Press

SALEM PRESS INC.
Pasadena, California   Hackensack, New Jersey

Some essays originally appeared in *Great Events from History: Science
and Technology Series* (1991), *The Twentieth Century: Great Events: 1900-2001*
(2002), *Great Events from History II: Science & Technology* (1991), *Great Events
from History: The Ancient World, Prehistory-476 c.e.* (2004), *Great Events from
History: The Middle Ages, 477-1453* (2005), *Great Events from History: The Renaissance & Early Modern Era, 1454-1600* (2005), *Great Events from History:
The Seventeenth Century, 1601-1700* (2006). New material has been added.

∞ The paper used in these volumes conforms to the American National Standard for Permanence of Paper for Printed Library Materials,
Z39.48-1992 (R1997).

Library of Congress Cataloging-in-Publication Data
Science and Scientists.
    p. cm. -- (Magill's choice)
    Includes bibliographical references and indexes.
    ISBN-10: 1-58765-302-8 (set : alk. paper)
    ISBN-13: 978-1-58765-302-5 (set : alk. paper)
    ISBN-10: 1-58765-305-2 (v. 3 : alk. paper)
    ISBN-13: 978-1-58765-305-6 (v. 3 : alk. paper)
    [etc.]
1. Discoveries in science. 2. Scientists--Biography. I. Salem Press. II.
Series.
    Q180.55.D57S29 2006
    509--dc22
                                                              2005030915

First Printing

PRINTED IN THE UNITED STATES OF AMERICA

Contents

Contents

Alphabetical List of Contents

Volume 1

VOLUME 2

Contents, xxvii

Volume 3

SCIENCE
AND
SCIENTISTS

Probability Theory

THE SCIENCE: Two mathematicians, Blaise Pascal and Pierre de Fermat, laid the foundations for probability theory when they responded to an inquiry about how to split the stakes from a game. Shortly thereafter, the first textbook on the subject was written by Dutch mathematician Christiaan Huygens.

THE SCIENTISTS:
Blaise Pascal (1623-1662), mathematician, theologian, and philosopher
Pierre de Fermat (1601-1665), mathematician and jurist
Christiaan Huygens (1629-1695), mathematician and physicist

CASTING THE DICE

Probability is the branch of mathematics that assesses how likely certain outcomes are when an experiment is performed. It entered the mathematical literature in the form of questions about games of dice, especially in the work of Gerolamo Cardano (1501-1576). These questions did not seem to have attracted much attention elsewhere, and Cardano's own work suffered from errors. It was not clear at the time how best to define even the basic notions on the basis of which to perform calculations involving events of chance.

Antoine Gombaud, the chevalier de Méré (1607-1684), was a French nobleman who was also a gambler, had investigated two problems. One was that of how to divide up the stakes in a game of dice when the game had to be broken off before it was finished. The other involved the likelihood of throwing a certain number of sixes in a certain number of throws of dice. De Méré knew how many tosses it would take to reach a 50 percent chance of at least one six landing face-up. He assumed that if he multiplied that number by six, he would have the number of tosses it would take for the likelihood of at least two sixes showing up to be more than 50 percent.

Experience showed that this was incorrect (and it had been the view of Cardano, although de Méré was unfamiliar with his work). Recognizing that he was out of his depth, de Méré turned to the eminent French mathematician Blaise Pascal. Pascal recognized the interest of the problems that had been proposed and initiated a correspondence in 1654 with perhaps the most accomplished mathematician of the period, Pierre de Fermat. Two of the world's greatest mathematicians thus turned their attention to a problem raised in the context of gambling. Both Pascal and Fermat were able to recognize the mathematical issues underlying the problem, and between them they created the theory of probability.

RECURSION

The nature of their arguments involved a precise analysis of the collection of possible outcomes at each stage of the games being played. Starting with a small number of principles, they could tackle both of the problems raised by de Méré by the use of a process now known as recursion. Recursion involves recognizing at certain stages of the game that the situation is exactly the same as it was at a previous turn and deriving from that recognition an algebraic equation that can be solved easily. Both Pascal and Fermat felt satisfied with the solutions that they obtained, although the absence of some of their correspondence does not provide a consistent basis for judging the generalizability of their arguments.

PASCAL'S TRIANGLE

One of the key ingredients to Pascal's solution was the triangle that bears his name. The triangle starts with a 1 at its apex, has two 1's in the next row, and continues with 1's at the ends of each row and interior elements obtained by adding up the two numbers immediately adjacent to it in the previous row. This particular triangle had been known for many years and went back at least to medieval Arabic mathematicians. What Pascal recognized was the way in which the numbers in a given row corresponded to the coefficients in expansions of a binomial expression, such as raising $(a + b)$ to the nth power. The amount of mathematical ingenuity that Pascal lavished on the triangle was impressive, but more surprising was the extent to which it enabled him to answer questions about probability.

Fermat's method of proceeding is less well documented, as is frequently the case with Fermat's work. His inclination was seldom to produce more than the details asked for in a problem rather than the method of proof. His willingness to calculate at length to enumerate all the possible outcomes of an experiment was the basis for his results, which agreed with those of Pascal.

Pascal had a religious conversion shortly after his correspondence with Fermat and gave up mathematics to a large extent. He made one further contribution to probability, however, which suggested the wider applications of their work. He framed an argument for belief in God that he suggested would be useful in arguing with those who needed to see everything put in terms of games and gambling. The argument used the idea of expectation and has remained an important contribution to philosophy.

Huygens's Theorems

The idea of "expectation" is connected with that of "average," and the rise of probability in the seventeenth century was perhaps connected with the availability of large quantities of data coming from national governments and other large bodies, such as municipalities. This notion provided the basis for the treatise on probability put together by the Dutch mathematician Christiaan Huygens (*Libellus de ratiociniis in ludo aleae*, 1657; *The Value of All Chances in Games of Fortune*, 1714). It is not clear how familiar he was with Fermat's and Pascal's work, but he did write the first systematic treatise on the rudiments of probability.

From a simple axiom he derived three theorems, and on the strength of those he explained the solutions to a sequence of problems, relying on the same sort of technique that had been used by Pascal. Where Pascal had used the combinatorial ideas embodied in his triangle, however, Huygens just lumbered through long computations. In a way, Huygens's work was a step back, but his casting the ideas of probability in a systematic form helped the subject to get something of a foothold among mathematicians.

Impact

Until the time of Pascal and Fermat, there was a tendency to appeal to arguments from inspiration and authority in many spheres. By the middle of the seventeenth century, the continued hostilities between Catholic and Protestant forces had cooled down to confrontations rather than conflict. In such a setting there was a call for the kind of argument that depended on something that could be accepted by both sides. Mathematics provided such a setting, and so there was a call for the ideas of probability in both Protestant and Catholic Europe.

Although the correspondence of Pascal and Fermat was not immediately available to subsequent mathematicians, the treatise by Huygens provided some impetus for further research. By the end of the century, there was an explosion of interest in probability, and a number of treatments of the basis of the subject took the place of Huygens's original work. Even in the middle of the eighteenth century, however, the leading authority on probability could look back on the subject as having been the creation of Pascal and Fermat. They had not been the first mathematicians to consider questions arising from games of chance, but they were the first to apply enough mathematical systematization to the subject to make sure that they did not fall into the traps that had bedeviled their predecessors and continue to afflict those who assess questions of probability without mathematics.

See also Abstract Algebra; Axiom of Choice; Bell Curve; Boolean Logic; Bourbaki Project; Calculus; Chaotic Systems; D'Alembert's Axioms of Motion; Decimals and Negative Numbers; Euclidean Geometry; Fermat's Last Theorem; Fractals; Game Theory; Hilbert's Twenty-Three Problems; Hydrostatics; Incompleteness of Formal Systems; Independence of Continuum Hypothesis; Integral Calculus; Integration Theory; Kepler's Laws of Planetary Motion; Linked Probabilities; Mathematical Logic; Pendulum; Polynomials; Russell's Paradox; Speed of Light.

FURTHER READING

Bernstein, Peter L. *Against the Gods: The Remarkable Story of Risk*. New York: John Wiley and Sons, 1996.

David, Florence N. *Games, Gods, and Gambling: The Origins and History of Probability and Statistical Ideas from the Earliest Times to the Newtonian Era*. New York: Hafner, 1962.

Gigerenzer, Gerd, et al. *The Empire of Chance: How Probability Changed Science and Everyday Life*. New York: Cambridge University Press, 1989.

Hacking, Ian. *The Emergence of Probability: A Philosophical Study of Early Ideas About Probability, Induction, and Statistical Inference*. Cambridge, England: Cambridge University Press, 1975.

Hald, Anders. *A History of Probability and Statistics and Their Applications Before 1750*. New York: John Wiley and Sons, 1990.

Maistrov, L. E. *Probability Theory: A Historical Sketch*. Translated and edited by Samuel Kotz. New York: Academic Press, 1974.

Todhunter, Isaac. *A History of the Mathematical Theory of Probability: From the Time of Pascal to that of Laplace*. Sterling, Va.: Thoemmes Press, 2001.

—*Thomas Drucker*

PSYCHOANALYSIS

THE SCIENCE: The foundations for psychoanalysis were laid by Sigmund Freud's *The Interpretation of Dreams*, in which he used dream analysis to introduce his influential theory that unconscious motives, molded from relationships in childhood, are basic to adult personality.

THE SCIENTISTS:

Sigmund Freud (1856-1939), Austrian neurologist who founded psychoanalysis

Josef Breuer (1842-1925), Austrian physician who worked with Freud
on hysteria

Erik Erikson (1902-1994), psychoanalyst who modified Freud's ideas

Hypnotism and Hysteria

Die Traumdeutung (1900; *The Interpretation of Dreams*, 1913) is widely
considered to be the greatest work of Sigmund Freud. This work is important because it introduced the core ideas of psychoanalysis. Although today the use of psychoanalysis as a therapy in mental health has changed
significantly, Freud's theory—that hidden, unconscious feelings and motives determine both the symptoms of mental patients and the normal
thoughts and deeds of everyday life—laid the foundations for modern
psychiatry and psychological therapies.

Even before he began his study of dreams, Freud, an Austrian neurologist, had already proved himself a capable medical researcher and produced several significant papers on neurological conditions. About 1885,
he was introduced to the study of hypnotism, and in the 1890's he worked
with Josef Breuer to develop a theory of hysteria. Breuer had called to the
attention of Freud the case of a young girl who suffered from apparent pa-
ralysis and psychic confusion. He noticed that if the girl were allowed to
give verbal expression to her fantasies, the symptoms tended to disappear.
Breuer also observed that, whereas the girl could not account for her symptoms in a conscious state, under hypnosis she well understood the connection between her symptoms and past experiences. From this case, Breuer
and Freud developed their theory: that hysteria is a condition that imitates
a physical or neurological disorder but for which no physical or neurological causes can be discovered. According to the theory, hysteria springs from
the repression of desired acts and can be cured only by a kind of catharsis
in which unconscious desires are rendered conscious and meaningful.

The Unconscious and the Unintentional

These studies in hysteria contained one basic idea that Freud was later
to develop in his theory of psychoanalysis: that a significant aspect of mental life was "unconscious." Inexpressible in words, the unconscious had indirect and sometimes perverse effects upon daily activity. In the 1890's,
Freud began to appreciate the general significance of his discovery. He began to analyze his own dreams and unintentional behavior. The unconscious, he realized, could be revealed in many ways other than hypnosis
and its significance was not limited to mental patients. *The Interpretation of*

Sigmund Freud. (Library of Congress)

Dreams was significant in that it introduced psychoanalysis not only as a treatment for hysteria but also as a comprehensive theory of human motivation and development.

Freud's work was distinguished both by the methodology he used to investigate dreams and by the meaning he assigned to dreams. He argued that the meaning of a dream is not to be discovered by some hidden external logic, but rather through a process of free association—by getting the *dreamer* to uncover its meaning. According to Freud, dreams are the protectors of sleep; they both express and censor unconscious desires, which are allowed free play once conscious mental activity is suspended. Thus the "manifest" dream—the dream that is remembered in a conscious state—is not the same as the "latent" dream-thought or desire, because this desire is often of such a nature (often sexual) that it conflicts with the requirements of society and the moral code that the individual self-imposes. The manifest dream partly censors this unconscious desire and at the same time expresses it in symbolic form.

Decoding Dreams

The decoding of such symbolism is the entrance into the complexes which, if not understood and rationally addressed, would lead to mental disorder. Several core themes of Freud's "dream book" became further elaborated in his later writings. The centrality of forbidden wishes (the "id") as modified and deflected by a "censor" remained one such continuing theme. This censor was in later work subdivided into the realistic controls of the conscious self: the "ego" as well as the less rational, moralistic restraints and demands of an internalized parental image, the "superego."

One core theme, the eroticized love for one's parent of the opposite sex and jealousy of one's same-sex parental rival, recurred in many dreams. This, later labeled the Oedipus complex, was considered by Freud to be basic to adult sexual identity and to neurosis. The mechanism of displaced symbolization which disguises forbidden dream wishes was later elaborated into Freud's many "mechanisms of defense."

MODIFICATIONS OF FREUDIAN THEORY

Not all of Freud's assumptions in 1900 have withstood the test of time. Freud's theory of motivation rested upon a hydraulic, tension-reducing analogy where such motives as sex and aggression would build up a sort of pressure that would demand some sort of release. The thrust of more recent psychology gives far more attention than did Freud to the joys of seeking out self-enhancing activities that often involve increased tension and excitement.

Major twentieth century psychoanalysts such as Erik Erikson give more emphasis than did Freud to the social interactions between parent and child quite apart from the sexual overtones of such relationships. Moreover, Freud's writings suffer in several ways from male biases characteristic of views of women prevalent in his time. Freud's account of little girls' family affections and jealousies was heavily flavored by an assumption of the biologically rooted inadequacy of females—an assumption that finds few defenders a century later. It has been charged that Freud too readily dismissed as fantasies reports by female patients of sexual abuse by trusted males.

FREUD ON SCIENCE

The work of Sigmund Freud has been the object of criticism for its now seemingly naive, as well as politically incorrect, approach to the human mind and personality. Many have condemned Freudian psychology, if not psychoanalysis as a whole, as "pseudoscience." No one, however, was more aware of the basis for this criticism than Freud himself, as he made clear in 1932:

In no other field of scientific work would it be necessary to insist upon the modesty of one's claims. In every other subject this is taken for granted; the public expect nothing else. No reader of a work on astronomy would feel disappointed and contemptuous of that science, if he were shown the point at which our knowledge of the universe melts into obscurity. Only in psychology is it otherwise; here the constitutional incapacity of men for scientific research comes into full view. It looks as though people did not expect from psychology progress in knowledge, but some other kind of satisfaction; every unsolved problem, every acknowledged uncertainty is turned into a ground of complaint against it.

Anyone who loves the science of the mind must accept these hardships. . . .

Source: Sigmund Freud, preface to *New Introductory Lectures on Psychoanalysis*, 1932. Reprinted in *A General Selection from the Works of Sigmund Freud*, edited by John Rickman (Garden City, N.Y.: Doubleday Anchor, 1957).

Other ideas found in Freud's "dream book" retain the vitality of having endured a century of research. Freud's thesis that dreams are meaningful clues to motives important in conscious, waking life is still treated with respect by many students of personality and biopsychology. With the discovery by twentieth century neuropsychologists—that dreaming episodes in sleep are accompanied by such distinctive neurophysiological signs as rapid eye movements—it became possible to study the nature of dreams with an objectivity greater than was possible for Freud. It appears that dreams are the result of random firing by neurons deep within the brain stem. Such dream episodes occur several times a night, and most are immediately forgotten. The few dreams that are remembered, however, may be precisely those that have personal significance.

Impact

Fundamentals of Freud's thought survive in psychoanalysis and in scientific psychology. Thousands of members of the International Psychoanalytic Association still practice their healing art. More important, basic Freudian ideas have become a vital, often unrecognized, part of mainstream psychology. Relationships between the quality of childhood-caretaker attachments and adult styles of relating to others form a popular focus for research in developmental psychology. The importance of implicit ("unconscious") adaptive styles, to cite another example, has become a key concern of cognitive psychology.

Post-Freudian art, literature, films, and television, no less than psychology, treat human emotions as subtle, complex, and often paradoxical, a view more consistent with Freud's portrayal of human nature than of prior nineteenth century conceptions of human rationality. Most of all, the study of the mind—which until 1900 was the domain of magic, religion, and speculative philosophy—has forever become the province of science. Without the stimulus of Freud's ideas, human understanding of life itself would not be at all the same.

See also Manic Depression; Pavlovian Reinforcement; REM Sleep; Split-Brain Experiments.

Further Reading

Erikson, Erik. *Childhood and Society*. New York: W. W. Norton, 1950.
Gay, Peter. *Freud: A Life for Our Time*. New York: W. W. Norton, 1988.
Jones, Ernest. *The Life and Works of Sigmund Freud*. 3 vols. New York: Basic Books, 1957.

Masson, Jeffrey M. *The Assault on Truth: Freud's Suppression of the Seduction Theory*. New York: Farrar, Straus, & Giroux, 1984.

Neu, Jerome, ed. *The Cambridge Companion to Freud*. New York: Cambridge University Press, 1991.

Nye, Robert D. *Three Psychologies: Perspectives from Freud, Skinner, and Rogers*. Pacific Grove, Calif.: Brooks-Cole, 1992.

Sulloway, Frank J. *Freud: Biologist of the Mind*. New York: Basic Books, 1979.

—Paul T. Mason and Thomas E. DeWolfe

Pulmonary Circulation

THE SCIENCE: Michael Servetus was the first person to publish his findings on how blood circulates from the heart, through the lungs, and then back to the heart, and how breathing has a function other than the cooling of the blood.

THE SCIENTISTS:

Michael Servetus (1511-1553), Spanish physician and church reformer

John Calvin (1509-1564), French Protestant theologian of the Reformation

William Harvey (1578-1657), English physician, first to establish firmly the function of the heart and describe the circulation of blood

Symphorien Champier (c. 1472-1539), French physician and founder of the medical faculty at Lyon, France

Johann Guenther von Andernach (c. 1505-1574), translator of Galen and a professor of medicine

THE RIGHT DIRECTION

During the Renaissance, the study of medicine relied primarily on the interpretation of the Greek and Latin texts of such figures as the Greek physician Hippocrates (c. 460-c. 377 B.C.E.) and the Roman physician Galen (129-c. 199). Although Servetus supported the medical views of his mentor Symphorien Champier, founder of the medical faculty at Lyon and a well-known Galenist, and while he expressed an acceptance of Galenism, his scholarly reflection allowed him to question strict Galenic ideas regarding the functions of the arterial and venous systems. In particular, he questioned the accepted notion that blood moved from the left to the right side of the heart through pores in the septum.

Servetus and Calvin: Scientific Persecution

Michael Servetus believed that the Church's teachings should be understandable to all the faithful, and his resultant theology was denied by Protestants and Catholics alike. In the period of the Reformation, such theological deviance could be life-threatening. Repudiated, Servetus fled to France.

He was welcomed to his new country by an arrest order from the Inquisition. Warned of the danger, he flirted with the idea of emigrating to the New World but instead enrolled at the University of Paris as Michel de Villeneuve. Later he moved to Vienne, just outside Lyons, where he was employed as editor and corrector for the firm of Trechsel. He quickly developed a friendship with Symphorien Champier, a local Humanist and doctor. In 1537, presumably on Champier's advice, Servetus returned to the University of Paris to study medicine. It was probably in Paris that Servetus made the medical discovery that is most commonly associated with his name: the concept of the pulmonary circulation of the blood. He supported himself by publishing medical pamphlets and lecturing on geography, but when he added astrology, he was soon in trouble again: He was brought before the Parlement of Paris to answer charges that included heresy. Fortunately he received a light sentence, but Servetus soon left Paris.

After two or three years at Charlieu, Servetus returned to Vienne, where spent the next twelve years (c. 1541-1553) working as physician and editor. There he initiated a correspondence with the Protestant reformer John Calvin—a former critic who did not welcome the communication. In 1546 Servetus sent a draft of what would become *Christianismi restitutio* (1553). In it, Servetus sought to restore the Church to its original nature and expanded on his idea that God is manifest in all things—skirting but not quite embracing pantheism. Calvin became increasingly exasperated and eventually stopped replying. Despite Servetus's requests, Calvin did not return his books and manuscripts, although he did send a copy of his own book, *Christianae religionis institutio* (1536; *Institutes of the Christian Religion*, 1561), which Servetus inscribed with sarcastic and critical annotations and returned.

Soon after the anonymous publication of *Christianismi restitutio* in January, 1553, Servetus was betrayed to the Inquisition. Calvin—who argued that Protestants should be no less ruthless than Catholics in the fight against heresy—had supplied evidence against him. Servetus escaped arrest but was found in August. Calvin worked to have Servetus prosecuted, and Servetus was condemned for heresy and sentenced to the stake. He was burned, dying in agony, on October 27, 1553. Calvin was never again challenged for control of Geneva.

Servetus formulated his concept of pulmonary circulation for the first time in 1546, contradicting Galen's misconceptions involving the functions of the lungs, and he accepted theories declaring the existence of pores in the septum separating right and left ventricles. Servetus stated that blood could pass from the right ventricle to the left only by means of the pulmonary artery and the lungs. This significant discovery in human physiology was incorporated into a manuscript of Servetus's, one on theological ideas called *Christianismi restitutio* (1553; partial translation, 1953), which was his final work. In the hope that his treatise would bring about

Michael Servetus. (National Library of Medicine)

a return to Christianity in its original form, Servetus sought but failed to find a willing publisher, primarily because his work incorporated heretical religious views involving the Trinity and opposition to the sacrament of infant baptism. Servetus, however, secretly agreed to print the manuscript in 1553 at his expense. A draft of the work was sent in 1546 to the Reformer of Geneva, John Calvin, who became Servetus's main enemy. The book was criticized vehemently from the moment of its release and its theories and claims led to Servetus's execution.

Undeniably, the small section of Servetus's ill-fated treatise that contained a detailed description of the pulmonary circulatory system constituted a significant anatomical breakthrough. Not only did Servetus describe the circulation of blood in the heart and the lungs accurately; his work heralded the declaration of the existence of general blood circulation, which was to be fully described seventy-five years later by the English physician William Harvey.

TRIPLE SPIRIT

Servetus's description of pulmonary circulation, however, was not an exercise in human anatomy alone. In addition, the work was theological. Servetus discussed the Holy Spirit, but he also argued, controversially, that there was a physiological basis to the principle of life. The principle of life was traditionally believed to be manifested in the form of a soul or vital spirit. Aristotle and Galen believed the heart to be the source of what was called animal heat, that blood circulated to warm the body, and that respiration's function was to cool the blood. Galenic thought, however, acknowledged that the vital spirit circulated in blood and originated in the liver. Servetus calculated that the soul of a human being was instilled during the first respiration at birth; the infant's first breath started the circulation of blood.

Servetus argued also for the existence of a "triple spirit" in humans: natural (specifically located in the liver and in the veins), vital (situated in the heart and arteries), and animal (seated in the brain and in the nerves). To explain how these parts of the spirit were joined together, Servetus reasoned that the vivifying factor resided in blood, which, because it constituted a moving component, connected all parts of the body. His idea was similar to the Hebrew conception that the soul resides in blood and originates from the "breath of lives." This conformed in large measure with Galen's teaching regarding the pneuma, that is, the soul or spirit.

Because Servetus had extensive knowledge of anatomical dissection, he could observe firsthand and thus describe the course of blood in the heart and the lungs precisely. Although he maintained the Galenic stance that the blood originated in the liver, Servetus amended Galen's claims that blood passes through orifices in the middle partition of the heart; Servetus had observed that in the heart, the primary movement of blood from right to left did not occur by way of the heart partition because it lacked orifices. This septum was not, according to Servetus, permeable to blood. Instead, he postulated that blood passed from the right ventricle to the left by means of a complex device, or communication joining the pulmonary artery with the pulmonary vein through a system of vessels by way of the lungs. Consequently, he figured that blood passed through the lungs to aerate, that is, to supply blood with oxygen through respiration; it was obvious to him that respiration was a physiological phenomenon. Yet he considered it also to be an aspect of divine process.

IMPACT

He showed that there were capillaries in the lungs and in the brain that join the veins with the arteries and perform special functions. This discovery of the pulmonary circulation of blood was a critical one whose effects are wide-ranging, and few figures in medicine can compare in stature and significance. In his final work, *Christianismi restitutio*, Servetus's description of the pulmonary circulation system linked oxygen, the air humans breathe, with life.

See also Blood Circulation; Human Anatomy.

FURTHER READING

Bainton, Roland H. *Hunted Heretic: The Life and Death of Michael Servetus, 1511-1553*. Boston: Beacon Press, 1960.

_____. "Michael Servetus and the Pulmonary Transit." *Bulletin of the History of Medicine* 7 (1938): 1-7.

Cunningham, Andrew. *The Anatomical Renaissance: The Resurrection of the Anatomical Projects of the Ancients*. Brookfield, Vt.: Ashgate, 1997.

Friedman, Jerome. *Michael Servetus: A Case Study in Total Heresy*. Geneva: Droz, 1978.

Goldstone, Lawrence, and Nancy Goldstone. *Out of the Flames: The Remarkable Story of a Fearless Scholar, a Fatal Heresy, and One of the Rarest Books in the World*. New York: Broadway Books, 2002.

Hillar, Marian, and Claire S. Allen. *Michael Servetus: Intellectual Giant, Humanist, and Martyr*. New York: University Press of America, 2002.

O'Malley, Charles Donald. *Michael Servetus: A Translation of His Geographical, Medical, and Astrological Writings, with Introductions and Notes*. Philadelphia: American Philosophical Society, 1953.

—*Karen R. Sorsby*

PULSARS

THE SCIENCE: In 1968, Antony Hewish and Jocelyn Bell announced the discovery of pulsars, a new class of star that provides the key to understanding supernovae and neutron stars.

THE SCIENTISTS:
Jocelyn Bell (b. 1943), graduate student in astronomy
Antony Hewish (b. 1924), radio astronomer and cowinner of the 1974 Nobel Prize in Physics

IDENTIFYING "SCRUFF"

The history of science is full of "accidental" discoveries that eclipse the original intention of the researcher and experiment. Jocelyn Bell's discovery of pulsars illustrates this phenomenon. (The word "pulsar" comes from "pulsating star.")

In 1965, astronomer Antony Hewish was constructing a new kind of radio telescope at the Mullard Radio Astronomy Observatory of the University of Cambridge. Although the telescope was designed to detect quasars (short for "quasi-stellar object"), quasars became the least significant part of the research. Bell, Hewish's graduate student, used the telescope to become the first person to identify a pulsar.

Radio astronomers locate objects in space by using radio telescopes to pick up the signals that these objects emit. While using these telescopes, astronomers may accidentally pick up something referred to as "noise." "Noise" generally means any unwanted radio signal or disturbance that interferes with the listener's ability to understand or use a desired incoming signal or to operate radio equipment. While learning how to use the new telescope, Bell discovered noise—something she called "scruff." The source of this unwanted signal noise was annoying, elusive, and invisible. Hewish's first thoughts were that the pulses were electrical noise within the instrument or perhaps some type of local noise such as ham-operated radios, automobile ignitions, or other electrical interference. Whatever the source, Bell was determined to find it.

Bell was able to distinguish whether the source was terrestrial or extraterrestrial using a simple but unique phenomenon: the difference between Earth time and star time. As the Earth makes its daily rotation about its axis, it also moves a little more than one degree around the Sun. Because of this, the Earth takes an extra four minutes to rotate in relation to the Sun, thus completing the twenty-four-hour day. In relation to the stars, however, a complete rotation of the Earth takes only twenty-three hours and fifty-six minutes. This is known as sidereal, or star, time. Bell observed the pulsating scruff over time and realized that it was synchronized not with Earth time but with sidereal time. This suggested an extraterrestrial origin for the scruff.

Jocelyn Bell. (The Open University)

LITTLE GREEN MEN OR WHITE DWARFS?

Because the pulses occurred with incredible precision (each pulse arrived at intervals of 1.3373011 seconds), Hewish and Bell had to consider the possibility that these regular pulses could be tangible evidence of alien intelligence. In good humor, they identified the source as LGM 1 (Little Green Men 1). The two astronomers were presented with a fascinating dilemma: If they announced the discovery without all the evidence and were proved wrong later, their research would become a textbook example of an improperly conducted scientific investigation. Yet, if these pulses were evidence of alien intelligence, the discovery was monumental. Hewish and Bell decided to attack the problem in the spirit and method of good science.

The LGM hypothesis faded; they renamed the source "CP 1919" (for Cambridge pulsar and its sky position) and turned their attention to describing the phenomenon. Hewish continued the survey over the Christmas vacation in 1967 and placed the raw data on Bell's desk. Upon her return, Bell began to analyze the chart and found a second source of pulses. Then, sources number three and four appeared. In the next two weeks, Bell was able to confirm that these were, indeed, independent sources. The nature of the source was still eluding Hewish, Bell, and other astronomers who had joined the search at Mullard Observatory. Nevertheless, Hewish and Bell announced their discovery in the journal *Nature* on February 24, 1968. They included a statement suggesting that these unusual sources might be traceable to white dwarf or neutron stars.

In publications later that year, Hewish seemed to favor the white-dwarf hypothesis. The editors of *Nature* seem to have favored the other option because on the cover of that issue were the words, "Possible Neutron Star." At this point, the problem of identifying the nature of the pulsating sources passed to the world community of scientists. The final linking of the pulsar with a rapidly rotating neutron star came with the combined work of astronomers Franco Pacini and Thomas Gold in 1968.

IMPACT

The announcement by Hewish and Bell triggered a flood of observational and theoretical papers on pulsars. In the following year, the list of pulsar locations grew to more than twenty-four; the current list includes more than four hundred. Pulsars were not discovered sooner because radio astronomers were using centimeter wavelengths to look at the sky, as opposed to Hewish's meter wavelengths. This was why the Hewish telescope was successful and Bell was able to resolve the pulses. For this and other outstanding work, Hewish shared the 1974 Nobel Prize in Physics with the British astronomer Sir Martin Ryle. It was the first awarded to astronomers. Hewish's award was based on his role in the detection of pulsars. Interestingly, Bell—the acknowledged discoverer—was not included in the Nobel recognition.

Bell, meanwhile, did not expect the instant celebrity status brought by the news of the discovery of pulsars, especially in the popular press. Bell quietly ended her observations, wrote her dissertation, and accepted a job in another field of research in another part of the country. The story of the pulsars became an appendix in her dissertation.

The history of pulsars appears to follow this sequence: First, a massive star explodes, causing a supernova; then, the core collapses, forming a neutron star. This star is rotating extremely rapidly, sending out beams of radio waves from two directions, or poles. These beams sweep through the universe much like the lights on top of a police cruiser. This becomes the scruff Bell identified on the radio telescope.

Pulsars stimulated further research on stellar evolution and its products such as white dwarfs, neutron stars, collapsars, frozen stars, and black holes. Observing binary neutron stars helps confirm Albert Einstein's general theory of relativity, the distortion of space-time near massive objects, and the existence of gravity waves. The way astronomers view the nature of the universe has changed because Jocelyn Bell persisted in understanding the nature of the scruff on her recording chart.

See also Big Bang; Black Holes; Brahe's Supernova; Cassini-Huygens Mission; Cepheid Variables; Chandrasekhar Limit; Copernican Revolution; Extrasolar Planets; Galactic Superclusters; Galaxies; Hubble Space Telescope; Neutron Stars; Quasars; Radio Astronomy; Radio Galaxies; Radio Maps of the Universe; Stellar Evolution; X-Ray Astronomy.

FURTHER READING

Astronomical Society of the Pacific. *The Discovery of Pulsars*. San Francisco: Author, 1989.

Greenstein, George. *Frozen Star*. New York: Freundlich Books, 1983.

_____. "Neutron Stars and the Discovery of Pulsars." *Mercury* 34 (March/April, 1985): 34-39, 66-73.

Hewish, Antony. "Pulsars, After Twenty Years." *Mercury* 38 (January/February, 1989): 12-15.

Hewish, Antony, et al. "Observation of a Rapidly Pulsating Radio Source." *Nature* 217 (February 24, 1968): 709-713.

—*Richard C. Jones*

QAFZEH HOMINIDS

THE SCIENCE: A team of scientists dated a modern-looking *Homo sapiens* fossil at ninety-two thousand years, more than doubling the length of time that modern humans had been known to exist.

THE SCIENTISTS:

Helène Valladas, French scientist at the Centre des Fables Radioactivités in France

Bernard Vandermeersch, French physical anthropologist and archaeologist

Dorothy Annie Elizabeth Garrod (1892-1968), English archaeologist and physical anthropologist

Sir Arthur Keith (1866-1955), eminent Scottish anthropologist

Theodore Doney McCown (1908-1969), American archaeologist and physical anthropologist

René Victor Neuville (1899-1952), French archaeologist

OUT OF AFRICA?

The origin of modern human beings has been a persistent and vexing problem for prehistorians. Part of the difficulty is that there is widespread disagreement over the relationship between modern humans and their closest extinct relative, Neanderthal man. Neanderthals differ from modern and some Archaic humans by the extreme robustness of their skeletons and by their heavy brow ridges, extremely large faces, and long and low skull caps. Neanderthals were once believed to have lived from approximately 100,000 to 50,000 years ago, while modern humans had been thought to have existed for 40,000 to 50,000 years. Although Neanderthals were originally assumed to have been the ancestors of modern humans, prehistorians now agree that they were too localized, too extreme, and too recent to have been forerunners of modern people.

Archaic fossil humans are less robust and make better candidates as ancestors of modern humans. These have been found in sub-Saharan Africa and also in the Middle East, where they overlap in both location and time with Neanderthals. No specimens of this type have been found in Europe.

Arguments regarding the origin of modern humans center on the issue of one ancestral group as opposed to many ancestral groups. One view assumes that modern humans evolved from many local Archaic types, including Neanderthal. A variation of this perspective holds that, while the ancestors of modern humans may have originated in one locality, they interbred with local peoples they met as they spread throughout the world. Both these views hold that this intermixture of genes explains the physical differences between modern populations.

Opposed to this view is the single-origin perspective, which maintains that earlier humans were replaced completely by physically and technologically more advanced members of a new group. The most favored homeland for this new and improved type is sub-Saharan Africa, where there are early examples of possible forerunners of modern humans, but where there is no known example of Neanderthals.

NEANDERTHALS IN THE LEVANT?

The region called Levant, which includes Israel, has been of considerable interest to holders of both of the theories just described because it forms a land bridge between Africa and the rest of the world. Any population moving from one region to the other had to pass through the Levant. This fact became particularly important when the first Neanderthal found outside Europe was discovered in Galilee in 1925.

Pursuing this lead, the English archaeologist Dorothy Annie Elizabeth Garrod excavated a series of caves on Mount Carmel, now in Israel, between 1929 and 1934. She was assisted by a young American, Theodore Doney McCown. In two of these caves, Tabun and Skhul, McCown discovered human remains with stone tools that had characteristics associated with Neanderthal remains.

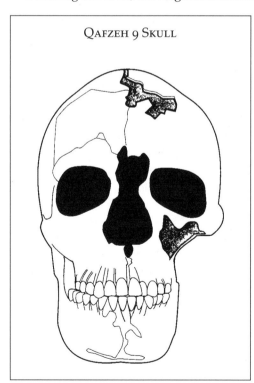

QAFZEH 9 SKULL

Back in England, McCown worked with the eminent Scottish anthropologist Sir Arthur Keith to analyze the bones. The fossils at Skhul, although Archaic, resembled modern humans in most characteristics; those at Tabun resembled Neanderthals, although these skulls' features were less exaggerated than those of classic Neanderthals.

Overlapping with Garrod's excavations at Mount Carmel were those by René Victor Neuville from the French consulate at Jerusalem. Neuville excavated the Qafzeh cave between 1933 and 1935, finding the remains of five individuals. Unfortunately, World War II (1939-1945) intervened, followed by the Israeli-Arab conflict of 1947. Neuville died without analyzing his material. From 1965 to 1975, the French archaeologist and physical anthropologist Bernard Vandermeersch continued Neuville's excavations, finding the remains of eight individuals who resembled the non-Neanderthals from Skhul.

The meaning of these remains was interpreted variously. McCown believed that the Mount Carmel population was in the process of diverging into two groups from a more generalized ancestor and that neither were modern humans. Others thought that the fossils represented a cross between Neanderthals and modern humans. A few others thought that Neanderthalls had been caught in the act of evolving into modern humans.

A major problem in making sense of the Levant fields lies in the inaccuracy of dates. Radiocarbon dating methods do not help because they are inaccurate for sites as old as Qafzeh, Tabun, or Skhul. Until recently, all that could be known was that humans of some sort had been in the Levant more than sixty thousand years ago and had lived there for an undetermined time.

Another method of dating, thermoluminescence, helped to clarify the dates. Thermoluminescent dating is used on objects such as pottery that were "fired," or heated during the time that they were used. When such objects are heated again in the laboratory, photons are released, producing thermoluminescence, or glow. The longer ago the object was fired, the more glow results. The greater the glow, the older the object. Thermoluminescence can be used to date much older material than can radiocarbon methods; unfortunately, it is not as accurate as radiocarbon.

The first objects to be dated by thermoluminescence in the Levant were burnt flints from the Neanderthal sites at Kebara. The dating was done by a French-Israeli team headed by Helène Valladas, with results being published in 1987. The Neanderthal site was dated at sixty thousand years, meaning that if the date is correct, Neanderthals were in the Middle East much later than had been thought.

In 1988, a team led by Valladas published a thermoluminescence date of ninety-two thousand years from Qafzeh. If this date is correct, then there were forerunners of modern humans living in the Levant twice as long ago as had been suspected. Furthermore, these individuals were there either before or at the same time as the Neanderthals.

IMPACT

Since thermoluminescence gives only a rough estimate, confirmation by another form of dating is desirable. In the meantime, there have been two dominant reactions by scientists. Those subscribing to the single-origin, out-of-Africa model see the Qafzeh date as confirmation of this hypothesis. Others, such as the American Milford Wolpoff, dispute this assessment. Wolpoff believes that Neanderthals contributed to the genetic makeup of modern Europeans. He points out that the late Neanderthals in Europe are more like modern Europeans in some respects than are the more modern-looking fossils from Skhul or Qafzeh.

The dates from Qafzeh have been adjusted and continue to raises many questions—particularly between those paleoanthropologists who believe that Neantherthals were basically replaced by modern humans and those who think the relationship between the species is more complex and may involve genetic exchange. Paleoanthropologists such as Richard Klein argue against genetic exchange; Wolpoff and his colleagues cite evidence for a more nuanced interpretation of the evidence that allows for an intermingling of gene pools.

See also *Australopithecus*; Cro-Magnon Man; Gran Dolina Boy; Human Evolution; Langebaan Footprints; Lascaux Cave Paintings; Lucy; Neanderthals; Peking Man; *Zinjanthropus*.

FURTHER READING

Garrod, D. A. E., and Dorothea Bate. *The Stone Age of Mount Carmel*. Vol. 1. Oxford, England: Clarendon Press, 1937.

Holloway, Ralph L. "The Poor Brain of *Homo sapiens neanderthalensis*: See What You Please." In *Ancestors: The Hard Evidence*, edited by Eric Delson. New York: Alan R. Liss, 1985.

Keith, Arthur. *An Autobiography*. London: Watts, 1950.

Klein, Richard. "Whither the Neanderthals?" *Science* 299, no. 5612 (March 7, 2003): 1525-1527.

McCown, Theodore D., and Arthur Keith. *The Stone Age of Mount Carmel*. Vol. 2 in *The Fossil Remains from the Levalloiso-Mousterian Levels*. Oxford, England: Clarendon Press, 1939.

Stringer, Christopher B., and Peter Andrews. "Genetic and Fossil Evidence for the Origin of Modem Humans." *Science* 239 (March 11, 1988): 1263-1268.

Trinkhaus, Eric, ed. *The Emergence of Modern Humans: Biocultural Adaptations in the Later Pleistocene*. Cambridge, England: Cambridge University Press, 1989.

Valladas, H., et al. "Thermoluminescence Dating of Mousterian 'Proto-Cro-Magnon' Remains from Israel and the Origin of Modern Man." *Nature* 331 (February 18, 1988): 614-616.

Wolpoff, M. *Paleoanthropology*. 2d ed. Boston: McGraw-Hill, 1999.

_____, et al. "Why Not the Neandertals?" *World Archaeology* 36, no. 4 (2004): 527-546.

—*Lucy Jayne Botscharow*

QUANTIZED HALL EFFECT

THE SCIENCE: The discovery of the quantized Hall effect—that an electric current traveling along a conductor will be bent toward one side when a magnetic force is imposed on the current—led to accurate measurements of certain fundamental constants of nature.

THE SCIENTISTS:

Edwin Herbert Hall (1855-1938), American physicist who discovered the Hall effect in 1879

Klaus von Klitzing (b. 1943), German physicist and winner of the 1985 Nobel Prize in Physics

Robert Betts Laughlin (b. 1950), American physicist who, among other scientists, provided explanations for the integral and fractional quantized Hall effect

SEMICONDUCTORS

With the explosive growth in microelectronics technology, semiconductors became some of the most studied of materials. The understanding of the basic microscopic phenomena in these crystals has advanced to the point where semiconductors are manufactured and manipulated on a molecular level. The properties of silicon, for example, are perhaps the best understood among solids. The measurement of these properties often reflects the complicated nature of the microscopic structure of these materials as well as the specifications of the particular sample. Given such detailed knowledge, the discovery of a novel fundamental phenomenon in semiconductors was startling.

The measurement of the quantized Hall effect depends only upon fundamental constants of nature and not on sample irregularities or impurities. The properties of a solid are especially dependent on a host of internal and environmental parameters, such as the geometry, temperature, purity of the sample, and the history of its preparation. It was surprising to find a manifestation of quantum mechanical behavior in macroscopic samples that is so distinct and precise.

THE CLASSIC HALL EFFECT

The Hall effect is a class of phenomena that occur when a material carrying current is subject to a magnetic field perpendicular to the direction of the current. As first observed in 1879 by the American physicist Edwin Herbert Hall, an electric voltage results in a direction perpendicular to both the current and the magnetic field. The ratio of this voltage to the current is the Hall resistance. In comparison, the normal electrical resistance is the ratio of the voltage in the direction of the current to the current. In the "classic" Hall effect, which occurs for a wide range of temperatures, the Hall resistance increases linearly with the strength of the magnetic field. The constant of proportionality depends on the individual characteristics of the sample and is a measure of the density of electrons that carry current.

The classic Hall effect is well described by what is called an "electron gas," where the motion of the conducting electrons of the solid is considered to be independent from each other and can freely wander within the crystal matrix. The case of the quantized Hall effect, however, requires a two-dimensional electron gas, where the electrons are confined to a plane of conduction. This is realized at the interface between a semiconductor and an insulator, where an electric field draws the semiconductor's electrons toward the two-dimensional interface. Temperatures of a few Kelvins are needed to keep the electrons stuck to the surface.

QUANTUM MECHANICS MEETS THE HALL EFFECT

It was demonstrated in 1966 that electrons confined to motion in such a plane, typically 10 nanometers thick, result in new quantum mechanical effects. The motion of the electrons is quantized, that is, their energies assume one of several evenly spaced discrete values. The number of electrons that can assume a particular energy, called a "Landau level," is proportional to the strength of the magnetic field. Increasing the field strength also increases the spacing between the Landau levels. At low temperatures, the electrons try to minimize their energy, and therefore, the Landau levels are filled sequentially by energy. The highest filled energy is called the "Fermi level." Raising the magnetic field effectively lowers the Fermi level, since more electrons can then be accommodated per Landau level. Alternatively, the Fermi level can be altered simply by changing the number of electrons.

Certain general aspects of the quantized Hall effect were, in fact, predicted in 1975 (five years prior to Klaus von Klitzing's experiments) by the Japanese theorists Tsuneya Ando, Yukio Matsumoto, and Yasutada Uemura of the University of Tokyo. They recognized that when every Landau level is either completely filled or completely empty, the electrical resistance should vanish. Under these conditions of "integral filling," the Hall resistance would be a certain ratio of fundamental constants divided by the number of filled levels and would be independent of the geometry. Unfortunately, the theory was only approximate and would not have been considered reliable for the actual experimental situation. The crucially important aspects of the extreme precision and the robustness of the effect under varying conditions were unforeseen. Also, in experiments as early as 1977 performed by von Klitzing's coworker Thomas Englert, slight plateaus in the Hall resistance were visible in some samples. These anomalous plateaus were considered unexplained by any published theories.

VON KLITZING'S RESEARCH

Von Klitzing's research through the 1970's included studies of silicon devices in high magnetic fields and under conditions of mechanical stress. In 1980, von Klitzing decided to investigate the anomalies in the Hall resistance. The high-quality samples he used were "metal oxide semiconductor field-effect transistors," or MOSFETs, constructed by his collaborators Gerhardt Dorda of the Siemens Research Laboratory in Munich and Michael Pepper of the University of Cambridge. A layer of insulating oxide is sandwiched between a metal strip, which provides a voltage potential, and the silicon, which supports the two-dimensional electron gas at its surface. The samples were typically about 0.4 millimeter long and .05 millimeter wide. By increasing the voltage on the metal electrode, more electrons could be drawn to the surface of the semiconductor, thereby raising the Fermi level.

Von Klitzing took his experiment to the High Field Magnetic Laboratory of the Max Planck Institute in Grenoble, France, to make measurements using their 20-tesla magnet, the magnetic field strength of which is roughly one million times stronger than the Earth's at ground level. Von Klitzing found for practically every sample that the Hall resistances were equal to the same fundamental ratio divided by integers to within a few percent, extending over well-developed plateaus in the variation of the

Fermi level. The subsequent high-precision results published were measured using the more stable 15-tesla magnet at the University of Würzburg. The accuracy improved to five parts per million, with the primary source of inaccuracy being the instability of the resistance standard. The ratio of the resistance at different plateaus, for example, was the ratio of integers to one part in thirty million. During the plateau regions, the electrical resistance fell very nearly to zero, ten times lower than any nonsuperconducting metal. Moreover, the resistivity continues to decrease as the temperature approaches absolute zero.

LACK OF RESISTANCE

The surprising features of the quantized Hall effect sent theoreticians into a flurry of activity. Impurities had been conventionally thought of as either trapping or deflecting electrons off their paths, giving rise to electrical resistance and causing variation in measurements from sample to sample. The seeming lack of involvement of impurities or defects was particularly enigmatic. In 1981, preliminary calculations by University of Maryland theorist Richard E. Prange suggested that although an electron can be trapped in a "localized state" around a defect in the crystal, under the condition that the Landau levels are integrally filled, the current lost to the trapped electron is exactly compensated by an increase in the velocity of electrons near the defect. The electrons move like a fluid, where flow speeds up around a barrier so that the total transported volume remains the same.

Theoreticians came to realize more generally that not only do impurities not cause resistance, but ironically they also are responsible for the plateaus in the Hall resistance as the magnetic field or the Fermi level is varied. The localized states act as a reservoir between Landau levels. As the Fermi level rises past complete filling of the conducting states of a given Landau level, only localized states are left to be filled up. The conducting electrons are effectively unaffected, giving rise to the constancy of the current as the Fermi level is varied.

IMPACT

The measurement accuracy of the fundamental ratio found in the quantized Hall resistance subsequently improved to one part in 108, and resulted in several immediate benefits. After a series of tests in independent laboratories was completed by the end of 1986, the quantized Hall effect was adopted as the international standard for resistance. The fine-structure constant, which is related to the fundamental Hall ratio by the speed of light, is a measure of the coupling of elementary particles to the electromagnetic field. Complementing high-energy accelerator experiments, the improved determination of the fine-structure constant provides a stringent test for theories of the fundamental electromagnetic interactions.

Soon after the integral quantized Hall effect was explained, a new "fractional" quantized Hall effect was discovered in 1982 by Dan C. Tsui, Horst L. Störmer, and Arthur Charles Gossard of Bell Laboratories. The type of sample they used for creating the two-dimensional electron gas, called a heterojunction, was made by a process called molecular beam epitaxy, where a layer of gallium arsenide positively doped with aluminum is grown onto a substrate layer of pure gallium arsenide. The gallium arsenide electrons are attracted toward the positively doped semiconductor and thus build up into a layer at the interface. The new device was a more perfect crystal and had better conduction properties, which were crucial for a successful observation of the fractional quantized Hall effect. In the fall of 1981, they brought their sample to the Francis Bitter Magnet Laboratory at MIT, where they used the 28-tesla magnet. They were searching at high fields and temperatures less than 1 Kelvin for an "electron crystal," where the electronic orbitals become arranged into a lattice. Instead, they found the same kind of plateaus and drops in the resistance observed in the integral quantized Hall effect, but occurring when only one-third or two-thirds of a Landau level is filled. Since then, many other fractions have appeared.

Theoretical investigations indicated that the observations could not be explained by an electron solid and demanded a radical description of the electronic behavior. In 1983, Robert Betts Laughlin gave a remarkable explanation in terms of a "quantum electronic liquid," in which the motions of the electrons are strongly affected by each other. The electronic liquid is incompressible: Rather than causing the density to increase, squeezing on the liquid causes a condensation of exotic fractional charges. These fractional charges play the role that electrons do in the integral Hall effect and so cause the plateaus at fractional values.

The impact on the field of physics reaches far beyond the accuracy of the measurement of the Hall resistance. Although the effect itself was not expected to be commercially significant, the MOSFET is essentially identical to components that may be important in the following generation of computers. Additionally, similarities emerged between the physical mechanisms of the fractional quantized Hall effect and those of high-temperature superconductors. Common features include a two-dimensional structure, low resistivity, and the collective motion of a macroscopic number of particles. The primary significance of the quantized Hall effects lies in revolutionizing and deepening an understanding of electronic properties of solids in high magnetic fields. For his work in this area, von Klitzing won the 1985 Nobel Prize in Physics.

See also Electron Tunneling; Quantum Mechanics.

FURTHER READING

Halperin, Bertrand I. "The Quantized Hall Effect: This Variation on a Classical Phenomenon Makes It Possible, Even in an Irregular Sample, to Measure Fundamental Constants with an Accuracy Rivaling That of the Most Precise Measurements Yet Made." *Scientific American* 254 (April, 1986): 52-60.
Klitzing, Klaus von. "The Quantized Hall Effect." *Reviews of Modern Physics 58* (July, 1986): 519-631.
MacDonald, Allan H., ed. *Quantum Hall Effect: A Perspective.* Boston: Kluwer Academic Publishers, 1989.
Schwarzschild, Bertram. "Von Klitzing Wins Nobel Physics Prize for Quantum Hall Effect." *Physics Today* 38 (December, 1985): 17-20.
—*David Wu*

QUANTUM CHROMODYNAMICS

THE SCIENCE: Murray Gell-Mann developed the theory of quantum chromodynamics to describe the characteristics of elementary particles called quarks.

THE SCIENTISTS:
Murray Gell-Mann (b. 1929), American physicist
Harald Fritzsch (b. 1943), German physicist
William Bardeen (b. 1941), American physicist

THE ATOM DIVIDED

It was not until the beginning of the twentieth century that it was discovered that atoms were not, in fact, indivisible; instead, they were actually made up of even smaller parts. In 1904, the first suggestion was made that the atom incorporated tiny subparticles called "electrons," which orbited a central core. In 1910, Ernest Rutherford, the English physicist, discovered the atom's core, which was called the "nucleus." Three years later, one of Rutherford's students, Niels Bohr, a Danish physicist, qualified the nature of the electron's orbit around the nucleus. By 1927, the problem of determining the atom's structure had been largely solved. A new science called "quantum mechanics" defined the atom's internal structure as consisting of tiny electrons orbiting a nucleus that contained an assortment of relatively heavy protons and neutrons. By 1930, a concentrated effort had been launched by a newly emerging branch of physics—particle physics—to probe the atom's secrets. Much evidence existed that there were smaller particles yet to be discovered within the atom's core.

SMASHING ATOMS

The first particle accelerator (atom smasher) was put into experimental use in 1932. The purpose of the accelerator is to cause atoms to collide with one another at extremely high speeds and break up into their elementary parts. Physicists then record the particles as they fly off in the collision. It was during a series of these particle accelerator experiments in the early 1960's that a physicist at the California Institute of Technology, Murray Gell-Mann, developed a series of brilliant postulations regarding the results of these particle accelerator experiments.

By late 1963, Gell-Mann had enough evidence to publish his theory that the nucleus of protons and neutrons was made up of even smaller particles. In reference to a passage in James Joyce's book *Finnegans Wake* (1939), Gell-Mann called these small pieces of protons and neutrons "quarks." He said he chose this name as "a gag . . . a reaction against pretentious scientific language." He published the first discussion of quarks in February, 1964. In 1969, he was awarded the Nobel Prize for his subatomic classification schemes.

QUARKS OF A DIFFERENT COLOR

Gell-Mann postulated six different kinds, or "flavors," of quarks (up, down, bottom, top, strange, and charm), each of which comes in three "colors" (red, green, or blue). The assignment of "colors" to quarks gave rise to a whole new branch of quantum physics—quantum chromodynamics (QCD).

There are no actual flavors in quantum mechanics (much less flavors defined as up, down, and so on), and likewise, at the subatomic level, there are no actual colors. These terms—"flavors" and "colors"—define the specific quantum characteristic of the elementary particle. Through classification and subclassification in the quantum chromodynamic nomenclature, the particles can be classed according to their characteristics and behavior.

Gell-Mann and his colleagues Harald Fritzsch and William Bardeen united the color concepts with the other quark ideas into a single formulation that united all the aspects of nuclear particles. Gell-Mann called presented this QCD theory in September, 1972. In the theory of QCD, the multicolored quarks are held together by a binding force called a gluon. This binding force is not only critical to any discussion of QCD but also fundamental to all of nature; it still drives the community of particle physics. Gluons make up what is called the "strong force," one of the four forces of nature.

IMPACT

QCD clarified a mixture of perplexing observations that had been compiled from numerous accelerator experiments. It enabled a clear understanding of some previously undefined observations. Furthermore, it so completely described the workings within the atomic nucleus that physicists were able to predict certain events before experiments were conducted—the ultimate validation of any theory.

QCD involves exceptionally difficult mathematics that strings together probabilistic mathematical events in a bewildering fashion. Because of this degree of difficulty, it becomes an intricate and enigmatic task to relate the data streaming in from particle accelerators to the field theory itself. Supercomputers have been employed to handle such processing, and all the final possible results from QCD have yet to be compiled.

Quantum chromodynamics is a scientific achievement that stands as a benchmark hypothesis on the landscape of physical theory. It fulfills the long-term dream of physicists of formulating a complete theory of the strong nuclear force (one of the four fundamental forces of nature) and the way in which it interacts with elementary particles at the atomic core. The final goal is to unify all four field theories of the forces of nature into a single grand unified theory of nature.

See also Quantum Mechanics; Quarks.

FURTHER READING

Crease, Robert P., and Charles C. Mann. *The Second Creation*. New York: Macmillan, 1986.

Gell-Mann, Murray. *The Quark and the Jaguar: Adventures in the Simple and the Complex*. New York: W. H. Freeman, 1994.

Hawking, Stephen W. *A Brief History of Time*. New York: Bantam Books, 1988.

Johnson, George. *Strange Beauty: Murray Gell-Mann and the Revolution in Twentieth-Century Physics*. New York: Alfred A. Knopf, 1999.

Pagels, Heinz R. *The Cosmic Code*. New York: Simon & Schuster, 1982.

_____. *Perfect Symmetry*. New York: Simon & Schuster, 1985.

Sutton, Christine. *The Particle Connection*. New York: Simon & Schuster, 1984.

Trefil, James S. *The Unexpected Vista*. New York: Charles Scribner's Sons, 1983.

—*Dennis Chamberland*

Quantum Mechanics

THE SCIENCE: In attempting to resolve anomalies in the traditional explanation of radiation emitted from certain heated objects, Max Planck restricted the object's "resonators" to discrete (or quantized) energies, an ad hoc solution that proved to have revolutionary implications.

THE SCIENTISTS:

Max Planck (1858-1947), German physicist, director of the Institute for Theoretical Physics of the University of Berlin, and winner of 1918 Nobel Prize in Physics

Albert Einstein (1879-1955), German-born American physicist and winner of the 1921 Nobel Prize in Physics

Rudolf Clausius (1822-1888), German physicist whose studies of the second law of thermodynamics deeply influenced the development of quantum theory

Ludwig Boltzmann (1844-1906), Austrian physicist whose statistical interpretation of the second law of thermodynamics influenced the development of quantum theory

Wilhelm Wien (1864-1928), German physicist whose studies of heat radiation influenced the development of quantum theory

Niels Bohr (1885-1962), Danish physicist who was director of the Institute of Theoretical Physics of the University of Copenhagen and winner of the 1922 Nobel Prize in Physics

THE PROBLEM OF BLACKBODY RADIATION

Toward the end of the nineteenth century, many physicists believed that all the principles of physics had been discovered and little remained to be done, except to improve experimental methods to determine known values to a greater degree of accuracy. This attitude was somewhat justified by the great advances in physics that had been made up to that time. Advances in theory and practice had been made in electricity, hydrodynamics, thermodynamics, statistical mechanics, optics, and electromagnetic radiation.

These classical theories were thought to be complete, self-contained, and sufficient to explain the physical world. Yet, several experimental oddities remained to be explained. One of these was called "blackbody radiation," the radiation given off by material bodies when they are heated.

It is well known that when a piece of metal is heated, it turns a dull red and gets progressively redder as its temperature increases. As that body is heated even further, the color becomes white and eventually becomes blue as the temperature becomes higher and higher. There is a continual shift of the color of a heated object from the red through the white into the blue as it is heated to higher and higher temperatures. In terms of the frequency (the number of waves that pass a point per unit time), the radiation emitted goes from lower to a higher frequency as the temperature increases, because red is in a lower frequency region range of the spectrum than is blue. These observed colors are the frequencies that are being emitted in the greatest proportion. Any heated body will exhibit a frequency spectrum (a range of different intensities for each frequency). An ideal body, which emits and absorbs all frequencies, is called a blackbody; its radiation when heated is called blackbody radiation.

The experimental blackbody radiation spectrum is bell-shaped, where the highest intensity—the top of the bell—occurs at a characteristic frequency for the material. The frequency at which this maximum occurs is dependent upon the temperature and

Max Planck. (Maison Albert Schweitzer/gift of Hans Kangro, AIP Emilio Segrè Visual Archives)

increases as the temperature increases, as is the case for any heated object. Many theoretical physicists had attempted to derive expressions consistent with these experimental observations, but all failed. The expression that most closely resembles the experimental curve was that derived by John William Strutt, third Baron of Rayleigh, and Sir James Hopwood Jeans. Like the experimental curve, it predicts low intensities for small frequencies; however, it never resumes the bell shape at high frequencies. Instead, it diverges as the frequency increases. Because the values of the theoretical expression diverge in the ultraviolet (high-frequency) region, it was termed the ultraviolet catastrophe. In other words, the Rayleigh-Jeans expression held that a body at any temperature would have its maximum frequency in the ultraviolet region. This was a clear contradiction to the experimental evidence.

PLANCK'S SOLUTION

On December 14, 1900, a soft-spoken, articulate professor presented a solution to the problem of blackbody radiation: Max Planck offered a mathematical exercise that averted the ultraviolet catastrophe. He explained why the heat energy added was not all converted to (invisible) ultraviolet light. This explanation was, to Planck, merely an experiment, a sanding down of a rough theoretical edge. It was that theoretical edge, however, that brought Planck before the august body of the German Physics Society. Six weeks earlier, he had done what he described as a "lucky guess." His discovery did not take place in a laboratory; it took place in his mind. Planck had introduced a mathematical construct into the Rayleigh-Jeans formula. Upon its use in the formula, however, Planck realized the significance of his mathematics. As Planck described, "After a few weeks of the most strenuous work of my life, the darkness lifted and an unexpected vista appeared."

Planck had discovered that matter absorbed heat energy and emitted light energy discontinuously. (Discontinuously means in discrete amounts of quantities, called quanta.) Planck had determined from his observations that the energy of all electromagnetic radiation was determined by the frequency of that radiation. This was a direct contradiction to the accepted physical laws of the time; in fact, Planck also had difficulty believing it.

QUANTUM THEORY

The work of Planck gave rise to quantum theory—a fundamental departure from the theory of light of his day. That theory stated that light waves behaved as mechanical waves. Mechanical waves, much like waves in a pond, are a collection of all possible waves at all frequencies, with a preponderance for higher frequency. In the mechanical wave theory, all waves appear when energy is introduced, and high-frequency waves

WHO REALLY DEVELOPED QUANTUM THEORY?

Until 1978, when historian of science Thomas Kuhn argued that Planck did not discover the quantization of radiant energy, a common interpretation was that Planck had understood these oscillators as actually emitting radiation in multiples of a definite energy that was proportional to its frequency. The proportionality constant, h, which came to be called Planck's constant, is an extremely small number that has the units of mechanical action. Planck saw this constant as a "mysterious messenger" from the microworld. He insisted that the "introduction of the quantum of action h" into physicists' theories about the atom "should be done as conservatively as possible." He knew that the classical wave theory of light had been shown to be true with many experimental observations, and he therefore wanted to preserve the continuous nature of radiation.

Nevertheless, Planck realized that he had accomplished something very important, because on December 14, 1900, when he first made his ideas on quantum theory public, he told his son Erwin that he had just made a discovery "as important as Newton's." On the other hand, he saw his greatest claim to fame in his radiation-law formula, since it agreed perfectly with energy distributions of radiations determined in laboratories for all wavelengths and temperatures.

The person who most profoundly understood the significance of Planck's work on quantum theory was Albert Einstein. He wholeheartedly embraced the idea of quantized energy and used it extensively in his work. For example, he used light quanta to explain the previously inexplicable photoelectric effect, an achievement for which he received the 1921 Nobel Prize in physics. Planck had won the 1918 Nobel Prize for "his discovery of energy quanta." By extending the discontinuity of energy to light as well as to the entire electromagnetic spectrum, and by his quantum studies of the interactions between light and matter, Einstein revealed the great power of the quantum idea.

In 1913, Niels Bohr developed his quantum theory of the hydrogen atom, using quantized electron energy states to account for the hydrogen spectrum. The full-fledged importance of the quantum idea became clear in quantum mechanics, developed in the 1920's by such eminent physicists as Louis de Broglie, Werner Heisenberg, and Erwin Schrödinger. So momentous was this new quantum theory that it has been the dominant theoretical tool for nearly a century, helping physicists and chemists to understand the microrealm of atoms and molecules.

fit more easily and therefore should be present in greater amounts. After empirical observation of blackbody radiation, Planck showed this to be incorrect.

He stated that light waves did not behave like mechanical waves. He postulated that the reason for the discrepancy lay in a new understanding of the relationship between energy and wave frequency. The energy either absorbed or emitted as light depended in some fashion on the frequency of light emitted. Somehow, the heat energy supplied to the glowing material failed to excite higher-frequency light waves unless the temperature of that body was very high. The high-frequency waves simply cost too much energy to be produced. Therefore, Planck created a formula that reflected this dependence of the energy upon the frequency of the waves. His formula said that the energy and frequency were directly proportional, related by a proportionality constant, now called Planck's constant. Higher frequency meant higher energy. Consequently, unless the energy of the heated body was high enough, the higher-frequency light was not seen. In other words, the available energy was a fixed amount at a given temperature. The release of that energy could be made only in exact amounts (later called quanta, or photons), by dividing up the energy exactly. Small divisions of the energy, resulting in large numbers of units, were favored over large divisions of small numbers of units. Lower frequencies (small units of energy) were favored over higher frequencies, and the blackbody radiation was explained.

Planck was quite reluctant to accept fully the discontinuous behavior of matter when it was involved with the emission of light or the absorption of heat energy. He was convinced that his guess would eventually be changed to a statement of real physical significance, for there was no way to see it, visualize it, or even connect it with any other formula. Yet, Planck's new mathematical idea had forced the appearance of a new and somewhat paradoxical physical picture.

IMPACT

Planck's simple formula started a furor in the world of physics, although he did not accept fully its conclusions. In fact, it was not accepted by most physicists at the time and was considered to be an ad hoc derivation. It was felt that in time a satisfactory classical derivation would be found. A few years later, however, the very same idea would be used in three different applications that would establish the quantum theory.

Albert Einstein—who understood the implications of Planck's work better than Planck himself—would use Planck's quantum theory to explain several other experimental oddities of the late nineteenth century. In 1905, Einstein explained the photoelectric effect using Planck's ideas. The phenomenon of the photoelectric effect is that light striking the surface of metals causes electrons to be ejected from that surface. Yet, it was also found that the phenomenon was frequency-dependent (not intensity-dependent). For example, a threshold frequency was needed to allow an electron to be ejected, not a threshold amount of energy caused by a high-intensity light source. Einstein showed that it was the frequency of the incident light that determined whether electrons would be ejected. He had used Planck's theory to explain that the threshold energy required to eject electrons was frequency-dependent. His correct theoretical explanation of the photoelectric effect won for him the 1921 Nobel Prize.

Two years later, in 1907, Einstein again utilized the quantum theory to explain one more "oddity." This explanation dealt with the capacity of objects to accept heat, their heat capacity. At the time, it was accepted that the heat capacity for any object at room-temperature conditions was constant. As the temperature of the object was decreased, however, the heat capacity was no longer at the classical value. In fact, these low-temperature heat capacities are quite contrary to classical theory. The classical result was for atoms vibrating about their equilibrium lattice positions. Einstein assumed that the oscillation of the atoms about their equilibrium lattice positions were quantized. In this instance, the mechanical vibrations of the atoms are subject to quantization. Therefore, in addition to electron oscillations and radiation itself, the motion of particles was found to be quantized.

In 1913, Niels Bohr used the ideas of the quantum theory to explain atomic structure. Bohr reasoned that the emission of radiation from excited atoms could only come about from a change in the energy of the electrons of those atoms. Yet, that radiation was found to be of discrete frequencies. Bohr determined that this implied that only certain quantized energy states were available to atoms and that the only means for change from one state to another was to have an exact (quantized) energy be emitted or absorbed. His theoretical explanation of the atom won for him the 1922 Nobel Prize.

Since these early successes of the quantum theory, the explanation of the microscopic has continued in quantum mechanics. The Planck quantum theory has become the basis for understanding of the fundamental theories of physics. In fact, all the "known" physics of the late nineteenth century has been shown to have a theoretical background based on the ideas of quantized light that converge to the classical theories at high temperatures and large numbers. Yet, the most basic understanding of matter and radiation physics has derived from the quantum theory.

See also Alpha Decay; Atomic Structure; Compton Effect; Electrons; Exclusion Principle; Grand Unified Theory; Heisenberg's Uncertainty Principle; Photoelectric Effect; Quantized Hall Effect; Quantum Chromodynamics; Schrödinger's Wave Equation; Superconductivity; Wave-Particle Duality of Light; X-Ray Fluorescence.

FURTHER READING

Brush, Stephen G. *Cautious Revolutionaries: Maxwell, Planck, Hubble.* College Park, Md.: American Association of Physics Teachers, 2002.

Cline, Barbara Lovett. *Men Who Made a New Physics.* Chicago: University of Chicago Press, 1987.

Duck, Ian. *One Hundred Years of Planck's Quantum.* River Edge, N.J.: World Scientific, 2000.

Heilbron, J. L. *The Dilemmas of an Upright Man: Max Planck as a Spokesman for German Science.* Berkeley: University of California Press, 1986.

Kuhn, Thomas S. *Black-Body Theory and the Quantum Discontinuity, 1894-1912.* New York: Oxford University Press, 1978.

—Scott A. Davis

QUARKS

THE SCIENCE: Physicists Murray Gell-Mann and George Zweig independently discovered that the disturbingly large number of so-called elementary particles could be effectively organized by assuming that they are composed of quarks.

THE SCIENTISTS:

Murray Gell-Mann (b. 1929), American particle physicist who won the 1969 Nobel Prize in Physics

George Zweig (b. 1937), particle physicist at the Swiss research institute CERN

Yuval Ne'eman (b. 1925), particle physicist at the Imperial College of London

THE TINIEST UNITS OF MATTER

At the end of the nineteenth century, the most elementary particles were thought to be atoms, the building blocks of molecules. During the early decades of the twentieth century, however, physicists discovered that the atom was composed of even smaller particles: protons, electrons, and neutrons. These "subatomic" particles were believed to be truly elementary.

Further developments in physics, however, began to reveal many more supposedly elementary particles. The number of such particles climbed from the familiar three in 1930 to more than one hundred by the 1980's. This large number was disconcerting to the physics community and became known as the "particle zoo." Physicists believed that all these particles could not be elementary.

THE EIGHTFOLD WAY

In 1964, in an attempt to simplify the field of elementary particle physics by identifying the smallest building blocks of matter, the existence of subatomic particles known as "quarks" was first postulated. It was suggested that many of the supposedly elementary particles were not elementary at all but were, instead, made up of these even smaller units.

As is often the case in science, this breakthrough was accomplished simultaneously by two scientists working independently: Murray Gell-Mann of the California Institute of Technology and George Zweig of Centre Européen de Recherche Nucléaire (CERN), a famous center for nuclear physics research in Zurich, Switzerland. Gell-Mann was led to the idea of quarks (he took the name from James Joyce's 1939 novel *Finnegans Wake*) by his analysis of mathematical and symmetrical relationships among some of the apparent groupings of the members of the particle zoo.

Together with Yuval Ne'eman, a particle physicist working in England, Gell-Mann had previously developed a way of organizing many of these particles into groups using a scheme called the "eightfold way." This scheme suggested that particles that appeared to be totally different were actually merely different versions of the same basic particle; the differences in appearance, Gell-Mann and Ne'eman theorized, were caused by different quantum numbers belonging to the various particles. (Quantum numbers specify certain physical properties of a particle, such as its charge and magnetic character.)

ACES, QUARKS, AND TRIPLETS

While Gell-Mann was developing the quark theory by analyzing the deep mathematical symmetries among the elementary particles, Zweig was led to the same ideas while trying to explain an experimental result.

THE COMPLEXITY OF MURRAY GELL-MANN

Murray Gell-Mann was considered a child prodigy and entered Yale University early in his teens, graduating in 1948 at the age of nineteen. Before the age of thirty, he had earned his doctorate in physics from the Massachusetts Institute of Technology, spent a year at the Institute for Advanced Study in Princeton, taught at the University of Chicago, and advanced to full professor at the California Institute of Technology, which he would make his academic home. In addition to winning the Nobel Prize in 1969, he received the Ernest O. Lawrence Memorial Award of the Atomic Energy Commission, the Franklin Medal of the Franklin Institute, the Research Corporation Award, and the John J. Carty medal of the National Academy of Sciences, the 1989 Science for Peace prize, and a host of other awards. He also served on the President's Committee of Advisors on Science and Technology (1994-2001).

This litany of honors and accomplishments—far from complete—belies the scope of Gell-Mann's brilliance, for he is far more than a theoretical physicist. If one were simply to list his contributions to that field—strangeness, the renormalization group, the V-A interaction, the conserved vector current, the partially conserved axial current, the eightfold way, current algebra, the quark model, quantum chromodynamics—they would rival those of giants like Isaac Newton and Albert Einstein. However, his passions extend far beyond physics, embracing linguistics, archaeology, history, psychology, and all matters of biological, cultural, and epistemological evolution. These interests are not restricted to theory; he is deeply concerned with environmental policy as well as world politics. His concept of the "effective complexity" of adaptive systems (which he defines as "the algorithmic information content—a kind of minimum description length—of the regularities of the entity in question") touches on nearly all areas of human thinking.

At the Santa Fe Institute, Gell-Mann has also drawn on the talents of linguists, archaeologists, anthropologists, and geneticists to explore the human language families and their relationships, in the hope of tracing language to its beginnings. For example, the project employs powerful computer programs to analyze different root words for the same meaning, taking into account their changes over time.

As physicist and fellow Nobel laureate Sheldon Glashow told biographer George Johnson: "Not only did Gell-Mann devise the lion's share of today's particle lore, but on first acquaintance you would soon learn . . . that he knew far more than you about almost everything, from archaeology, birds and cacti to Yoruban myth and zymology."

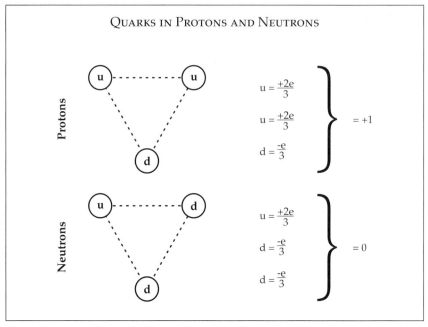

QUARKS IN PROTONS AND NEUTRONS

Protons

u u

d

$u = \frac{+2e}{3}$

$u = \frac{+2e}{3}$

$d = \frac{-e}{3}$

$\left.\begin{array}{c} \\ \\ \end{array}\right\} = +1$

Neutrons

u d

d

$u = \frac{+2e}{3}$

$d = \frac{-e}{3}$

$d = \frac{-e}{3}$

$\left.\begin{array}{c} \\ \\ \end{array}\right\} = 0$

Quarks do not exist as free particles. Rather, they combine to form subatomic particles with integer charges, such as the proton (with a charge of +1) and the neutron (with a charge of 0).

Zweig noticed that when a certain supposedly elementary particle, the pi-meson, disintegrated into other particles, it did so in an unusual way. To explain this unexpected finding, Zweig suggested that the pi-meson was composed of two parts, with individual properties (known as "strange-ness") that were transmitted separately to the decay components. To make this scheme work theoretically, Zweig found it necessary to postulate that many of the particles were constructed from an underlying triplet of parti-cles that he called "aces." It was soon determined that Zweig's "aces" were the same as Gell-Mann's "quarks," and "quarks" became the accepted term for the new fundamental triplet.

The nature of the individual quarks, however, was controversial. Gell-Mann believed that they might be purely mathematical entities that would never be detected in the way that other particles are—by the trail of bub-bles that they leave in specially designed chambers used to chart their paths. Zweig, however, believed that quarks should be physically observ-able.

To resolve this dilemma, researchers began experiments to search for individual quarks. They searched in accelerators, in cosmic rays, in chunks of normal matter, even in oysters, but to no avail. Quarks were nowhere to be found, suggesting that Gell-Mann had been correct.

In 1968, however, experiments at the new Stanford Linear Accelerator Center (SLAC) showed that electrons bouncing off protons were recoiling in a way that suggested they were hitting something hard and small inside the proton. Nevertheless, no quarks were observed directly and individually, despite much effort to find them.

Physicists came to believe that quarks are bound together very tightly by particles called "gluons." The strength of the binding is so great that quarks can never be separated from one another. Thus, quarks are fundamental building blocks of larger particles, but they exist only in combination with other quarks and can never be observed independently.

IMPACT

The significance of the idea of quarks lies in the central role that it plays in the theories developed by the physics community to explain the interactions among the various kinds of natural forces. One of the deepest mysteries in twentieth century physics, for example, was the nature of the nuclear force that holds protons tightly packed together in the nucleus of an atom. Since protons are all positively charged, they experience a powerful electrical repulsive force that should push them apart. Yet there is a force—the "strong force"—that holds them together. The source of the strong force was a mystery until the development of the quark theory.

Perhaps the most significant accomplishment of the quark theory will be its role in the development of a "unified field" theory, the hunt for a grand unified theory. Scientists hope that such a theory will one day unify the explanations of all the various natural forces under a single theoretical umbrella. Some unity has been achieved via the creation of various grand unified theories. These theories show that the electromagnetic, weak, and strong forces are similar in that each has a "force carrier" called, respectively, the "photon," the "intermediate bosons," and the "gluon." Physicists are searching for the "graviton," the postulated carrier of the gravitational (and fourth) force. It has been suggested that all of these forces may have emerged from a single force during the first few moments of the "big bang" that created the universe. If an understanding of gravity can be incorporated into one of the grand unified theories, then scientists will have shown how all the forces are merely different manifestations of a single original force. The quark theory is an indispensable part of this grand search.

See also Quantum Chromodynamics.

FURTHER READING

Carrigan, Richard A., and W. Peter Trower, eds. *Particle Physics in the Cosmos*. New York: W. H. Freeman, 1989.

_____. *Particles and Forces: At the Heart of the Matter*. New York: W. H. Freeman, 1990.

Crease, Robert P., and Charles C. Mann. *The Second Creation: Makers of the Revolution in Twentieth Century Physics*. New York: Macmillan, 1986.

Glashow, Sheldon, with Ben Bova. *Interactions: A Journey Through the Mind of a Particle Physicist and the Matter of This World*. New York: Warner Books, 1988.

Ne'eman, Yuval, and Yoram Kirsh. *The Particle Hunters*. New York: Cambridge University Press, 1986.

Riordan, Michael. *The Hunting of the Quark*. New York: Simon & Schuster, 1987.

Trefil, James S. *From Atoms to Quarks: An Introduction to the Strange World of Particle Physics*. New York: Charles Scribner's Sons, 1980.

—*Karl W. Giberson*

QUASARS

THE SCIENCE: Maarten Schmidt recognized that previously mysterious "quasi-stellar objects," or "quasars," must be very luminous, very distant objects in space.

THE SCIENTISTS:

Maarten Schmidt (b. 1929), astronomer who discovered the nature of quasars

Allan Rex Sandage (b. 1926), astronomer at the Mount Wilson and Mount Palomar Observatories

Thomas A. Matthews (b. 1924), astronomer who specialized in identification of strong radio sources

Cyril Hazard (b. 1925), radio astronomer at Jodrell Bank Radio Observatory

GOD'S FINGER

Between 1960 and 1963, radio astronomy (which tries to detect objects in space by intercepting the radio signals that these objects emit) was faced with a major puzzle: Several sources of radio waves had been identified in the sky that seemed to have no visible counterpart. Whereas most previously studied radio sources were either peculiar galaxies or nearby gas clouds, this new type of source seemed to have no such identity. Known by their numbers in the massive Third Cambridge Catalog of Radio Sources (the 3C catalog), the best studied of these sources were labeled "3C 48," "3C 286," and "3C 196," referring to their positions in space. These sources would later be called "quasars." Diameters were measured for these objects in 1960 at the giant Jodrell Bank radio telescope in England by Cyril Hazard and his colleagues, who found them to be surprisingly small.

Intrigued by the peculiarities of these objects, Allan Rex Sandage took photographs of these radio sources in September, 1960, using the Palomar Observatory's 508-centimeter telescope, then the largest in the world. Sandage and Thomas A. Matthews, an expert at identifying radio sources, studied the photographs but found nothing that resembled a normal radio galaxy or gas cloud. They noticed, however, that the photograph of the area at the position of 3C 48 included a star with a peculiar feature: A faint wisp of light seemed to be pointing at it, "as if," as Sandage excitedly exclaimed, "God's finger were pointing to the true radio source." At the next opportunity, in October of that year, Sandage obtained a spectrum of this star and measured its colors. (A spectrum is a picture of the light of a star that has been spread out into all of its different colors, like a rainbow. This picture is marked by vertical black lines. Astronomers can use the black lines in a star's spectrum to discover the composition and velocity of the star.)

Sandage's spectrum of 3C 48 was extremely puzzling. As he explained to the members of the American Astronomical Society at their December, 1960, meeting, the spectrum resembled nothing that had been seen before. Instead of a bright continuum of light of different colors with various dark lines, the spectrum of 3C 48 had a weak continuum with broad, fuzzy lines. The most puzzling feature, however, was that none of these lines corresponded with any elements that were known to be contained in stars; the lines were completely unidentifiable. The only thing that the colors definitely showed was that it was a very hot object, with a temperature on the order of 100,000°.

THE MOON AS YARDSTICK

A major breakthrough occurred in 1962, when Hazard and his collaborators used the Parkes Radio Telescope in Australia to make a high-precision measurement of the position of 3C 273. As seen from Parkes, the

Moon happened to pass directly over the position of this object, and therefore a careful measurement of the time of its disappearance and later reappearance gave a very accurate measurement of its location, as the position of the moving Moon was known very accurately. When they compared the radio position with optical photographs of that part of the sky, the astronomers found that the position corresponded exactly with that of a fairly bright star. (The apparent brightness of this "star," which is the brightest quasar in the sky, is approximately six hundred times fainter than the faintest star visible without a telescope. Other quasars have since been found that are ten thousand times fainter.)

When this quite positive identification was announced, Maarten Schmidt of the California Institute of Technology decided to use the Palomar telescope to obtain a photograph and a spectrum of the "star." The photograph showed a bright stellar object with a faint wispy structure to one side, "pointing" toward the other object, much like what was found by Sandage next to 3C 48. The spectrum looked much like that of 3C 48, but with the broad emission lines in entirely different places. This remarkable fact threatened to confound the situation even more, until Schmidt had a brilliant insight as he examined the spectrum. He realized that the object's spectral lines would make sense if they were, in fact, normal lines of common elements but redshifted greatly to longer than usual wavelengths. (A source of light that is moving rapidly away from an observer will have all of its light shifted in wavelength to redder, longer wavelengths, by an amount that depends upon its velocity.) If he identified four of the lines as being caused by hydrogen gas—the most common element in the universe—then he found that the object must be moving away from Earth at about 48,000 kilometers per second. Comparing this information with the rate of speed at which the universe is believed to be expanding, it was possible to measure reliably the distance to 3C 273: about two billion light-years.

Impact

It took nearly two decades of study for astronomers to accept that quasars are the extremely

Maarten Schmidt. (California Institute of Technology)

bright centers of normal galaxies. The mechanism that explains their nearly incredible amounts of energy must be gravitational collapse, since astronomers know of no other way to explain them. These galaxies probably draw material (mostly hydrogen gas) from neighboring galaxies. This material has fallen into the center of the object, where it has collapsed to form a very massive black hole (so called because the gravity of a black hole prevents light or any signals from escaping). The black hole is not seen; however, the newly captured gas that is falling toward it heats up to extreme temperatures (hundreds of thousands of degrees) and emits huge amounts of light (brighter than one trillion suns) as it is pulled toward oblivion.

Schmidt's discovery of the nature of the quasars in 1963 led to new and surprising insight into Earth's cosmic environment. Quasars represent the oldest and most distant objects that can be viewed, because they are seen as they appeared billions of years ago, when the universe was young. Thus, their properties can tell astronomers something about the properties of the universe long ago; for example, galaxy collisions and interactions were far more common then (billions of years ago), even more common than simple models would predict. The quasars also tell what happens when massive objects collapse, forming black holes at the centers of galaxies.

See also Big Bang; Black Holes; Brahe's Supernova; Cassini-Huygens Mission; Cepheid Variables; Chandrasekhar Limit; Copernican Revolution; Extrasolar Planets; Galactic Superclusters; Galaxies; Hubble Space Telescope; Neutron Stars; Pulsars; Radio Astronomy; Radio Galaxies; Radio Maps of the Universe; Stellar Evolution; X-Ray Astronomy.

FURTHER READING

Burbidge, Geoffrey, and Margaret Burbidge. *Quasi-stellar Objects*. San Francisco: W. H. Freeman, 1967.

Hodge, Paul. *Galaxies*. Cambridge, Mass.: Harvard University Press, 1986.

Kahn, F. D., and H. P. Palmer. *Quasars: Their Importance in Astronomy and Physics*. Manchester, England: Manchester University Press, 1967.

Rowan-Robinson, Michael. *Cosmology*. Oxford, England: Clarendon Press, 1977.

Sciama, D. W. *Modern Cosmology*. Cambridge, England: Cambridge University Press, 1971.

Weedman, Daniel W. *Quasar Astrophysics*. Cambridge, England: Cambridge University Press, 1986.

—Ronald B. Guenther

RADIO ASTRONOMY

THE SCIENCE: An antenna set up to detect the causes of interference with radio transmission detected the first radio signals from outside the solar system.

THE SCIENTISTS:
Karl Jansky (1905-1950), American radio engineer
Albert Melvin Skellett (b. 1901), American radio technician and astronomer
Grote Reber (1911-2002), American radio engineer who became the first radio astronomer

A STRANGE HISS

In 1928, Karl Jansky was hired by Bell Telephone Laboratories as a radio engineer. His first assignment was to investigate the causes of interference with transatlantic radio-telephone transmissions. This investigation required a sensitive antenna whose frequency response and sensitivity were very stable—ideal characteristics of any radio telescope. The device Jansky built, which was called a "Bruce array," consisted of two parallel frameworks of brass tubing; one frame was connected to a receiver, and the other acted as a signal reflector. The antenna was mounted on four Model-T Ford wheels and rotated every twenty minutes on a circular track. The antenna, which was about 20 meters long and 4 meters in width and height, was nicknamed the "merry-go-round."

Jansky discovered that there were three kinds of signals that his instrument could detect. Nearby thunderstorms created infrequent but powerful radio bursts. Distant thunderstorms created a weak but steady signal as their radio signals were reflected off the ionosphere, an electrically conducting layer in the upper atmosphere. The third signal, which created a steady hiss in receivers, was at first a mystery. Even though this signal was not a serious problem for radio reception, Jansky continued his efforts to identify the source. The signal varied in intensity in a daily cycle, and Jansky initially suspected that it might originate with the Sun. The problem with this theory was that the signals reached their highest point a few minutes earlier each day.

Jansky, who was unfamiliar with astronomy, did not appreciate the significance of this observation, but a friend of his, Albert Melvin Skellett, did. The Earth takes 23 hours and 56 minutes to rotate with respect to the stars. Because the Earth moves in its orbit by about one degree per day, it takes an extra four minutes to complete a rotation with respect to the Sun. The signals were following sidereal (star) time; that is, they came from a source that was fixed with respect to the stars. In 1933, after a full year of observations, Jansky published his estimate of the source's location: in the southern part of the Milky Way galaxy in the direction of Sagittarius. In 1935, after additional analysis, Jansky reported that signals originated from all along the Milky Way.

COSMIC SIGNALS

Once Jansky understood the nature of the cosmic signals, he found that he was completely unable to detect the Sun, which he found quite puzzling. Jansky happened to be observing at a time of minimal sunspot activity. If he had observed at a time of great sunspot activity (at "sunspot maximum"), his equipment should have detected solar radio emissions. Had he observed at sunspot maximum, however, the upper atmosphere would have been nearly opaque at the wavelengths he studied, and he probably would not have detected radio waves from the Milky Way. Jansky realized that if he could not detect the Sun, the signals from the Milky Way were not likely to originate in the stars. He suggested that the radio signals originated from interstellar dust and gas instead, a suspicion that has proved to be correct.

Jansky's observations were described in a front-page article in *The New York Times* on May 5, 1933, and a national radio program broadcast a few seconds of cosmic radio noise. Nevertheless, the discovery had little significance for practical communications. Jansky proposed the construction of a 30-meter dish antenna to study the cosmic signals in greater detail, but his employers, believing that such investigations were more appropriate for academic researchers, turned down the proposal. Jansky went on to other areas of communications research and received a commendation for his work on radio direction finders during World War II. He had always been in poor health, and he died in 1950 at the age of forty-four, as radio astronomy was beginning to flourish.

One of the few people who had sufficient knowledge of both astronomy and radio to take advantage of Jansky's work was Grote Reber, who realized that investigating celestial radio sources would require completely different equipment from that Jansky had used. In 1937, Reber built a parabolic reflecting antenna with a diameter of 10 meters, which was used to make maps of the sky by aiming the parabolic dish at different elevations and letting the Earth's rotation sweep the antenna across the field of view.

IMPACT

The discovery of cosmic radio signals led to the field of radio astronomy—the first time astronomers used any part of the electromagnetic spectrum other than the range of frequencies containing visible and infrared light. This new tool allowed astronomers for the first time to investigate the universe without having to depend on optical telescopes: Because radio waves penetrate cosmic dust and gas clouds, which block visible light, these radio waves could be used to map the structure of the Milky Way galaxy. In the years since Jansky and Reber's work, radio astronomy has discovered great explosive bursts in other galaxies, some of which emit so much energy that their cause is became a focus of scientific investigation. Radio astronomy also discovered pulsars, as well as the faint background radiation that most astronomers consider to be the echo of the big bang. Astronomers were unprepared for the discovery that the universe could look so different at radio wavelengths.

Perhaps the most important effect of radio astronomy was to teach astronomers that every part of the electromagnetic spectrum reveals new phenomena and new types of celestial objects. The result has been new areas of investigation, including X-ray and ultraviolet astronomy.

See also Cosmic Microwave Background Radiation; Ionosphere; Isotopes; Pulsars; Quasars; Radio Galaxies; Radio Maps of the Universe; Very Long Baseline Interferometry; Wilkinson Microwave Anisotropy Probe.

FURTHER READING

Hey, J. S. *The Evolution of Radio Astronomy.* New York: Science History, 1973.

Reber, Grote. "Radio Astronomy." *Scientific American* 181 (September, 1949): 34-41.

Spradley, Joseph L. "The First True Radio Telescope." *Sky and Telescope* 76 (July, 1988): 28-30.

Sullivan, Woodruff T., III. "A New Look at Karl Jansky's Original Data." *Sky and Telescope* 56 (August, 1978): 101-105.

_____. "Radio Astronomy's Golden Anniversary." *Sky and Telescope* 64 (December, 1982) 544-550.

Verschuur, Gerrit L. *The Invisible Universe Revealed: The Story of Radio Astronomy.* New York: Springer-Verlag, 1987.

—*Steven I. Dutch*

RADIO GALAXIES

THE SCIENCE: Martin Ryle's interferometric radio telescope detected and provided details on the structure of the first identifiable radio galaxy.

THE SCIENTISTS:

Karl Jansky (1905-1950), American radio engineer who detected the first cosmic radio waves

Grote Reber (1911-2002), American radio engineer who became the first radio astronomer

Sir Martin Ryle (1918-1984), English radio physicist and astronomer

Francis Graham Smith (b. 1923), English radio physicist and astronomer

Walter Baade (1893-1960), German American astronomer

Rudolf Minkowski (1895-1976), German American physicist

TUNING IN TO EXTRATERRESTRIAL RADIO

The initial measurements of cosmic radio emission by the American radio engineers Karl Jansky and Grote Reber between 1932 and 1940 showed reasonable similarity between the universe as it was revealed by radio waves and as it was seen by optical telescopes. This led many astronomers to conclude that most, if not all, celestial radio emissions came from interstellar gas, which was evenly distributed throughout the universe. Until the post-World War II period, the greatest drawback of the new discipline of radio astronomy was its limited accuracy in determining the celestial position and structural detail of an object detected from radio signals. A higher degree of accuracy was necessary in order to give optical astronomers a small enough "window" in which to look for an object discovered by radio telescopes.

Immediately following World War II, J. S. Hey used receivers from the Army Operational Radar Unit to study some of the extraterrestrial radio emissions reported earlier by Jansky and Reber. In 1946, Hey and his colleagues reported an observational discovery of particular import: that a radio source in the constellation Cygnus varied significantly in strength over very short time periods. In contrast to Reber, who had concluded that interstellar hydrogen between the stars was the source of all celestial radio signals, Hey argued that the fluctuations were too localized for interstellar gas and suggested instead the existence of a localized, starlike object.

In Australia, a similar group was formed under the direction of J. L. Pawsey. In 1946, the Australian group verified Hey's observations of a localized radio source in Cygnus, using one of the earliest radio interferometers. An interferometer is a series of radio telescopes connected over a wide area. The resolution, or receiving power, of this array is equal to that of a single radio telescope with a diameter equal to the distance between the farthest single radio telescopes. More "radio stars" were discovered by Pawsey's group in June, 1947, using an improved Lloyd's interferometer developed by L. McCready, Pawsey, and R. Payne-Scott. Incorporating an antenna mounted on top of a high cliff, this interferometer was able to use the ocean to reflect radio waves.

RYLE'S PYROMETER

Almost simultaneously, using a different type of interferometer, Sir Martin Ryle at Cambridge found another intense localized radio source in the constellation Cassiopeia. Ryle and others had extensive wartime experience in developing airborne radar detectors, radar countermeasures, underwater sonar arrays, and signal detection and localization equipment. Ryle was joined in 1946 by Francis Graham Smith, who helped him rearrange his interferometer to cover a wider receiving area. Ryle's cosmic radio "pyrometer" was used successfully in June, 1946, to measure a large sunspot.

Hey and his colleagues remained unable to determine the accurate position of their radio source to better than 2°. The successful resolution of solar sunspots of small diameter, however, suggested to Ryle that improvements to the radio telescope were possible. In 1948, Ryle, Smith, and others made the first detailed radio observations of Cygnus A using an improved version of their pyrometric radio telescope. Ryle and Smith subsequently published an improved position for Cygnus A. Ryle and Smith's measurements included the discovery of short-period radio bursts, which they (incorrectly) used to argue that the ultimate radio source must be a radiating star of some unrecognized type.

EXTRAGALACTIC RADIO SOURCES

In 1949, astronomers decided to establish the locations of Cygnus A and Cassiopeia by constructing a very large interferometric radio telescope, with a maximum separation of 160 kilometers. In 1951, Ryle developed a new "phase-switching" receiver based on 1944 sonar detection efforts. Phase switching permits the radio receiver to reject sources with large size and thus improves the receiver's ability to emphasize and locate weaker sources. After completing the prototypes, in 1951 researchers again made measurements of Cygnus A and Cassiopeia. They discovered that the radio sources were clearly not stars and that the Cygnus A source was actually two distinct sources. Ryle's phase-switched records were such improvements that his colleagues compiled the first radio object catalog, which listed more than fifty cosmic radio sources.

By late 1951, Smith had further localized the coordinates of Ryle's two radio stars, reducing the original error windows of Hey and others by a

RYLE'S RADIO TELESCOPES

In 1964, Sir Martin Ryle implemented his first history-making radio telescope, the "one-mile telescope," using the principle of aperture synthesis. Aperture synthesis uses small telescope dishes to produce the angular resolution of a much larger telescope dish. A telescope's angular resolution is its ability to distinguish between two relatively close point sources of radiation. The method of aperture synthesis keeps one or more small dishes fixed and moves one or more other dishes over Earth's surface, comparing the phases of the radiation collected by the fixed and movable dishes. Ryle's instrument was unique because he accounted for the effects of Earth's rotation in moving the array of dishes and provided a baseline for the angular resolution which was a large fraction of Earth's diameter.

The telescope had a resolution superior to that of existing instruments, and it could detect much fainter sources, including quasars. Quasars are among the most distant, and therefore youngest and most powerful, objects in the universe. They may represent the stage galaxies go through before their radio radiations subside and they become visible at optical wavelengths. With Allan Sandage, Ryle developed a technique for identifying quasars at optical wavelengths, based on the fact that they emit much more ultraviolet energy than single, normal stars.

Ryle's survey of radio sources showed, as suspected, that the number of faint sources per unit volume of space increased with distance—but far more rapidly than anticipated. This finding supported the big bang theory of the universe, which then was in conflict with the prevailing "steady state" theory. Over time, Ryle's evidence was bolstered by other research, and today the big bang theory is dominant.

In 1974, Ryle won the Nobel Prize in Physics for construction and use of the five-kilometer telescope. Like the one-mile telescope, it consisted of a linear, east-west array of dishes, some fixed and some movable, all carried by Earth's rotation. The resolution of this instrument was, remarkably, one second of arc. The new telescope detected fainter and more distant galaxies with greater precision, allowing many more objects to be used for statistical studies and the radio and optical components of the most distant objects to be matched. The five-kilometer telescope was also used to study individual stars in the Milky Way that were just being born. Ryle's telescopes thus opened a new window on the universe, revealing previously undetectable objects and insights.

factor of sixty. Smith then approached the director of the Cambridge Observatory to seek visual identification of the two radio sources. While part of the Cassiopeia source, a supernova remnant, was found in 1951, the poor atmospheric observing conditions in England prevented the Cambridge observers from making a complete identification. Shortly thereafter, Smith sent his data to Walter Baade and Rudolf Minkowski of the Palomar Observatory in Southern California.

The objects of Baade and Minkowski's visual search were discovered only after many difficulties; they were hidden among many other stars and faint galaxies. In 1952, Baade wrote to Smith that the result of his visual search was puzzling. He had found a cluster of galaxies, and the position of the radio source coincided closely with the position of one of the brightest members of the cluster. Moreover, the source seemed to be receding at a high velocity. These findings suggested that the source of the radio transmissions was outside the galaxy.

At first, there was notable skepticism over the notion of extragalactic radio sources. Because of this climate of disbelief, Baade and Minkowski did not publish their results until 1954. Subsequent research, however, confirmed the extragalactic location of these objects, forcing cosmologists to find a place for these sources in their theories of the universe.

IMPACT

Perhaps the most decisive radio data came from Ryle and Scheuer in 1955. They found that the most "normal" galaxies emit radio signals comparable in intensity to those emitted by the Milky Way galaxy. Nevertheless, there were many other galaxies—many not different in their optical appearances from normal galaxies—which are much more powerful sources; these became known as "radio galaxies." By the time of the 1958 Solvay Conference on the Structure and Evolution of the Universe, the existence of at least eighteen extragalactic radio objects had been confirmed, opening a new era in cosmology.

See also Cosmic Microwave Background Radiation; Ionosphere; Isotopes; Pulsars; Quasars; Radio Astronomy; Radio Maps of the Universe; Very Long Baseline Interferometry; Wilkinson Microwave Anisotropy Probe.

FURTHER READING

Baade, W., and R. Minkowski. "Identification of the Radio Sources in Cassiopeia, Cygnus A, and Puppis A." *Astrophysical Journal* 119 (1954): 206-214.

Pawsey, I. L., and R. N. Bracewell. *Radio Astronomy*. Oxford, England: Clarendon Press, 1955.

Ryle, M., and F. G. Smith. "A New Intense Source of Radio Frequency Radiation in the Constellation Cassiopeia." *Nature* 162 (1948): 462.

Sullivan, Woodruff T., ed. *Classics in Radio Astronomy*. Boston: D. Reidel, 1982.

_____. *The Early Years of Radio Astronomy. Reflections Fifty Years After Jansky's Discovery*. New York: Cambridge University Press, 1984.

Verschuur, Gerrit. *The Invisible Universe: The Story of Radio Astronomy*. New York: Springer-Verlag, 1974.

—*Gerardo G. Tango*

RADIO MAPS OF THE UNIVERSE

THE SCIENCE: Grote Reber built the first radio telescope and used it to record the first radio contour maps of the Milky Way, establishing the foundations of a new type of astronomy.

THE SCIENTISTS:

Grote Reber (1911-2002), American radio engineer and amateur astronomer

Karl Jansky (1905-1950), American radio engineer

Sir William Herschel (1738-1822), German English musician and astronomer

Harlow Shapley (1885-1972), American astronomer

OPENING AN INVISIBLE WINDOW

Grote Reber's recording of the first radio contour maps of the universe was a new and unexpected application of radio technology. Reber's work opened the invisible window of radio frequencies, allowing astronomers to see new features of the universe.

Sir William Herschel was one of the first astronomers to recognize the true nature of the dense band of stars across the sky called the Milky Way. From counting stars in various directions in the Milky Way, he concluded in 1785 that the vast majority of stars are contained within a flattened disk shape, forming an island universe or galaxy in space, with the solar system reduced to a tiny speck in the vast universe of stars. Early in the twentieth century, Harlow Shapley was able to use the 254-centimeter Mount Wilson telescope to show that the Milky Way galaxy is far larger than any previous estimate, and that the Sun is far from the galactic center, which he located in the direction of the constellation Sagittarius.

In 1932, Karl Jansky reported his accidental discovery of radio waves from space. Using a rotating array of antennas sensitive to 15-meter radio waves, he detected a steady hiss whose emission corresponded to the daily motion of the stars. He concluded that he was receiving cosmic radio waves from beyond the solar system. He was able to identify the source of the most intense radiation in the direction of Sagittarius, suggesting that it came from the center of the Milky Way galaxy. He also showed that weaker radio waves came from all directions in the Milky Way and suggested that their source was in the stars or in the interstellar matter between the stars.

THE LONE RADIO ASTRONOMER

Jansky's work was so unrelated to traditional astronomy that no professional astronomer followed it up. As a young radio engineer at the Stewart-Warner Company in Chicago, Reber read Jansky's papers and began to plan how he could measure the detailed distribution of the radiation intensity throughout the sky at different wavelengths. In 1937, he built a 9.4-meter parabolic reflecting dish in his yard, mounted so that it could be pointed in a north-south direction; scanning west to east would result from the Earth's rotation. For ten years, he operated this radio telescope in Wheaton, Illinois, as the only radio astronomer in the world.

Grote Reber. (National Radio Astronomy Observatory, operated by Associated Universities, under contract with the National Science Foundation)

As the Milky Way crossed the meridian late at night, Reber measured the increasing intensity of the cosmic radio waves. He published his initial results in the February, 1940, *Proceedings of the Institute of Radio Engineers*, where he noted that the intensity of the radiation was too low to come from stars, as Jansky had proposed, but suggested the possibility of radiation from interstellar gases.

In 1941, Reber began a complete sky survey with an automatic chart recorder and more sensitive receiving equipment. The recording pen would slowly rise and fall as the reflecting dish rotated with the Earth. After collecting approximately two hundred chart recordings, he plotted the resulting radio contours on the two hemispheres of the sky. The resulting radio maps, published in the *Astrophysical Journal* in November, 1944, revealed interesting details: The greatest radio intensity was coming from the center of the galaxy, in Sagittarius; less intense radio waves were coming from the constellations Cygnus and Cassiopeia. More important was his recognition that radio waves could penetrate the interstellar dust that obscures much visible light in the Milky Way.

Reber's last observations in Wheaton were made from 1945 to 1947. The resulting radio maps, published in the *Proceedings of the Institute of Radio Engineers* in October, 1948, now revealed two noise peaks in the Cygnus region, later identified as a radio galaxy (Cygnus A) and a source associated with a spiral arm of the Milky Way (Cygnus X). An intensity peak in Taurus was later identified with the eleventh century supernova remnant in the Crab nebula, and another in Cassiopeia matches the position of a seventeenth century supernova explosion. These results were the beginning of many important discoveries in the field of radio astronomy.

IMPACT

Reber's pioneering work and resulting radio maps led to a growing interest in radio astronomy and many unexpected discoveries with radio telescopes of increasing sophistication and size. In 1945, a graduate student at the University of Leiden, in the Netherlands, Hendrik Christoffel van de Hulst, predicted that neutral hydrogen should emit 21-centimeter radio waves. By 1949, the Harvard physicist Edward Mills Purcell had begun to search for these radio waves with Harold Irving Ewen, a graduate student who was sent to confer with Reber on techniques in radio astronomy. Ewen and Purcell developed special equipment and by 1951 had succeeded in detecting the predicted 21-centimeter radio waves. A group headed by Dutch astronomer Jan Hendrik Oort then began a seven-year collaboration with Australian radio astronomers to map the spiral arms of the Milky Way galaxy.

In 1960, two radio sources were identified with what appeared to be stars, but each emitted much more radio energy than Earth's sun or any other known star. Four of these "quasars" (quasi-stellar radio sources) had been discovered by 1963. At distances of billions of light-years, these objects would have to be more than one hundred times brighter than entire galaxies and would appear to be some kind of highly energetic stage in the early formation of a galaxy.

Another dramatic event in radio astronomy occurred in 1967, when Jocelyn Bell, a graduate student in radio astronomy at Cambridge, discovered pulsars. These are believed to be fast-spinning "neutron stars" with high magnetic fields that produce a rotating beam of radio emission. A pulsar in the Crab nebula was later identified with the collapsed core of the supernova remnant that had appeared on Reber's radio maps.

Perhaps the most important discovery in radio astronomy was the 1965 detection of cosmic microwave background radiation by radio astronomers Arno Penzias and Robert Woodrow Wilson. Using a 6-meter horn antenna at the Bell Telephone Laboratories in Holmdel, New Jersey, they found an unexpected excess of steady radiation with no directional variation. This matched current predictions of cosmic radiation from a primeval fireball in the "big bang" theory. Thus, radio astronomy provided confirmation of the creation and expansion of the universe.

See also Cosmic Microwave Background Radiation; Ionosphere; Isotopes; Pulsars; Quasars; Radio Astronomy; Radio Galaxies; Very Long Baseline Interferometry; Wilkinson Microwave Anisotropy Probe.

FURTHER READING

Abell, George O., David Morrison, and Sidney C. Wolff. *Exploration of the Universe*. 5th ed. Philadelphia: Saunders College Publishing, 1987.
Hey, J. S. *The Evolution of Radio Astronomy*. New York: Science History Publications, 1973.
Spradley, Joseph L. "The First True Radio Telescope." *Sky and Telescope* 76 (July, 1988): 28-30.
Sullivan, W. T., III, ed. *The Early Years of Radio Astronomy*. Cambridge, England: Cambridge University Press, 1984.
Verschuur, Gerrit L. *The Invisible University Revealed*. New York: Springer-Verlag, 1987.

—*Joseph L. Spradley*

RADIOACTIVE ELEMENTS

THE SCIENCE: Frédéric Joliot and Irène Joliot-Curie used alpha particles from polonium to bombard aluminum and create phosphorus 30, an artificial nucleus that is radioactive.

THE SCIENTISTS:
Irène Joliot-Curie (1897-1956), French physicist who shared the 1935
    Nobel Prize in Chemistry with her husband, Frédéric Joliot
Frédéric Joliot (1900-1958), French physicist who shared the 1935 Nobel
    Prize in Chemistry with his wife, Irène Joliot-Curie

NEUTRONS AND POSITRONS

In 1930, a German team of scientists reported that beryllium bombarded by alpha particles emitted a new sort of penetrating radiation. In France, the husband-wife team of Irène Joliot-Curie and Frédéric Joliot confirmed the German results and in the process came close to proving the existence of the neutron, of which the new radiation was composed. This particle, which has no charge but has the same mass as the proton, joins the proton to form the nucleus of the atom. British physicist James Chadwick won a Nobel Prize in Physics for making the actual discovery of the neutron.

In 1932, the Joliot-Curies, studying cosmic radiation in the high Alps, observed the positron, a particle with the same mass as the electron but with a positive rather than a negative charge. They failed to follow up their observation, and that same year, an American physicist, Carl David Anderson, identified the positron, using equipment similar to that used by the Joliot-Curies.

ALPHA BOMBARDMENT

By early 1933, the Joliot-Curies were using alpha particles produced from polonium to bombard boron, beryllium, fluorine, aluminum, and sodium. After the bombardment, these elements emitted neutrons and both positrons and electrons. The accuracy of the results of their experiments was questioned at the Seventh Solvay Conference, which was attended by most of the major physicists in Europe. They returned to Paris with damaged pride and a new determination to prove conclusively that neutrons and positrons were emitted at the same time from their irradiated targets.

To conduct the necessary experiments, they were forced to modify their experimental apparatus. Until now, the Geiger counter, which detected radioactivity, had been automatically turned off when the radioactive source (the source of the alpha particles) was removed. In the new arrangement, it would be left on after the source was removed. With this arrangement, they noticed that aluminum continued to emit positrons for some time after the removal of the radioactive source. This meant that the aluminum target had been made artificially radioactive by bombardment with alpha particles.

The Joliot-Curies were certain that they had produced artificial radioactivity. In order to place their discovery beyond doubt, they needed to separate chemically the source of the new radioactivity and to demonstrate that it had nothing to do with the original aluminum target. On January 15, 1934, friends, including Irène's famous mother, Marie Curie, received frantic telephone calls from the young researchers and rushed to the laboratory. From makeshift apparatus scattered in apparent disarray over several tables, the Joliot-Curies bombarded aluminum with alpha particles and separated from the irradiated samples an isotope

Irène Joliot-Curie. (Library of Congress)

of phosphorus with a half-life of only three minutes and fifteen seconds. Marie Curie, who was dying of the leukemia produced by her lifetime work with radioactivity, was handed a tiny tube containing the first sample of artificially produced radioactivity. Her face expressed joy and excitement. Other colleagues filled the room with lively discussions.

The Joliot-Curies soon repeated their experiments with boron and magnesium, producing still other sources of artificial radioactivity. They promptly sent off a report of their discovery to the scientific press. Its publication opened a floodgate of new experiments on the transmutation of nuclei, which led directly to the discovery of "nuclear fission" five years later.

IMPACT

The report of the discovery of artificial radioactivity was published early in 1934 and in 1935 earned for the husband and wife a shared Nobel Prize in Chemistry. The scientific community almost immediately recognized the discovery as equal to that of the neutron or the positron. Physicist Enrico Fermi and his group in Rome quickly noted that neutrons were more effective in producing artificial radioactivity than the alpha particles used in the original experiments. The entire community, including the Joliot-Curies, began to study artificial radioactivity produced by bombarding different elements with neutrons. Studies on uranium in Rome, Berlin, and Paris led to confusing results, which were finally interpreted as nuclear fission in 1939. (Nuclear fission is the splitting of an atomic nucleus into two parts, especially when bombarded by a neutron. When the nuclei of uranium atoms are split, great amounts of energy are released.)

SCIENCE AND ROMANCE

Irène Joliot-Curie was the daughter of the legendary Marie Curie, the physical chemist who twice won the Nobel Prize, and a member of the French scientific elite by birth as well as a brilliant physicist in her own right. As a teenager, she had worked alongside her mother, using X-ray equipment to treat soldiers wounded during World War I. She published her first paper in physics in 1921, and in 1932 she succeeded her mother as director of the Radium Institute.

Frédéric Joliot grew up in a middle-class family and attended the École de Physique et de Chimie Industrielle de la Ville de Paris rather than one of the prestigious French universities. Because of his unquestionable ability as an experimenter, he was recommended to Marie Curie as an assistant by a close friend, French physicist Paul Langevin. He joined the laboratory at the end of 1924 and gradually acquired the necessary degrees. Because of his background, he found it difficult to break into the inner circle of French science, despite his personal charm and ability.

The outgoing, charming, handsome Frédéric fell in love with the quiet, capable, socially awkward Irène. In 1926, they were wed, beginning a very happy marriage and an extremely successful scientific collaboration. During the first four years of the 1930's, they embarked upon a remarkable series of experiments in nuclear physics that led to the creation of radioactive elements in the laboratory. Their achievement led directly to nuclear fission, achieved in 1939.

The Joliot-Curies continued to lead a rich family life hampered only by poor health caused, in Irène's case, by her early work with large amounts of radioactive materials. Frédéric Joliot was now accepted as a member of the French scientific elite and not as an upstart who had married Madame Curie's daughter. As World War II loomed on the horizon, Frédéric Joliot was drafted into the military. Recognizing the possibility of a nuclear fission bomb, he took steps to secure uranium for France and began negotiations for a large supply of "heavy water" located in Norway. With the Nazis closing in, he and his colleagues smuggled the heavy water to Britain and hid the uranium in Morocco just ahead of Adolf Hitler's advancing troops. During the war, Joliot used the prestige of his Nobel Prize to conceal his activities in support of the French Resistance. Thus, in addition to being a major scientific contributor in his own right, Joliot helped tilt the war in the direction of the Allies.

See also Amino Acids; Atomic Nucleus; Atomic Structure; Atomic Theory of Matter; Boyle's Law; Buckminsterfullerene; Carbon Dioxide; Chlorofluorocarbons; Citric Acid Cycle; Definite Proportions Law; Isotopes; Liquid Helium; Neutrons; Osmosis; Oxygen; Periodic Table of Elements; Photosynthesis; Plutonium; Spectroscopy; Vitamin C; Vitamin D; Water; X-Ray Crystallography; X-Ray Fluorescence.

FURTHER READING

Biquard, Pierre. *Frédéric Joliot-Curie: The Man and His Theories.* Translated by Geoffrey Strachan. New York: Paul S. Eriksson, 1966.
Goldsmith, Maurice. *Frédéric Joliot-Curie.* London: Lawrence and Wishart, 1976.
Jungk, Robert. "An Unexpected Discovery." In *Brighter Than a Thousand Suns,* translated by James Cleugh. New York: Harcourt Brace, 1958.
Opfell, Olga S. "Irene Joliot-Curie." In *The Lady Laureates: Women Who Have Won the Nobel Prize.* Metuchen, N.J.: Scarecrow Press, 1978.
Rhodes, Richard. "Stirring and Digging." In *The Making of the Atomic Bomb.* New York: Simon & Schuster, 1986.

—*Ruth H. Howes*

RADIOMETRIC DATING

THE SCIENCE: Bertram Borden Boltwood pioneered the radiometric dating of rocks, leading to the use of nuclear methods in geology and establishing a new chronology of Earth.

THE SCIENTISTS:
Bertram Borden Boltwood (1870-1927), the first American scientist to study radioactive transformations
Ernest Rutherford (1871-1937), English physicist who won the 1908 Nobel Prize in Chemistry
William Thomson, Lord Kelvin (1824-1907), English physicist

NOT ENOUGH TIME

Radioactivity is the property exhibited by certain chemical elements that, during spontaneous nuclear decay, emit radiation in the form of alpha particles, beta particles, or gamma rays. Related to this property is the process of nuclear disintegration. In this process, an atomic nucleus, through its emission of particles or rays, undergoes a change in structure. To take an example, the presence of helium in rocks is the result of radioactive elements in the rocks emitting alpha particles during disintegration.

Bertram Borden Boltwood was fascinated by the theory of radioactive disintegrations proposed in 1903 by McGill University scientists Ernest Rutherford and Frederick Soddy. According to that theory, radioactivity is always accompanied by the production of new chemical elements on an atom-by-atom basis. In 1904, Boltwood impressed Rutherford by demonstrating that all uranium minerals contain the same number of radium atoms per gram of uranium. This confirmation of the theory of radioactivity marked the beginning of a close collaboration between the two scientists.

The significance of Boltwood's work can be seen in the light of a chronological controversy raging at that time between geologists and physicists. It had been accepted generally that the Earth, at some time in its history, was a liquid ball and that its solid crust was formed when the temperature was reduced by cooling. The geological age of Earth was, thus, defined as the period of time necessary to cool it down from the melting point to its present temperature. Using these guidelines, several estimates of that time were made by the famous physicist Sir William Thomson, Lord Kelvin, who in 1877 claimed that the age of the Earth was probably close to 20 million years and certainly not as large as 40 million. His calculations were mathematically correct but did not take into account radioactivity, which had been discovered the year before.

Geologists believed that 40 million years simply was not enough time to create continents, to erode mountains, or to supply oceans with minerals and salts. Studies of sequences of layers (stratigraphy) and of fossils (paleontology) led them to believe that the Earth was older than 100 million years, but they were not able to prove it.

Earth Clocks

Boltwood and Rutherford proved that geologists were right. This came as a by-product of their research on the nature of radioactivity. Rutherford knew that helium was always present in natural deposits of uranium, and this led him to believe that in radioactive minerals, alpha particles somehow were turned into ordinary atoms of helium. Accordingly, each rock of a radioactive mineral is a generator of helium. The accumulation of the gas proceeds more or less uniformly so that the age of the rock can be determined from the amount of trapped helium. In that sense, radioactive rocks are natural clocks.

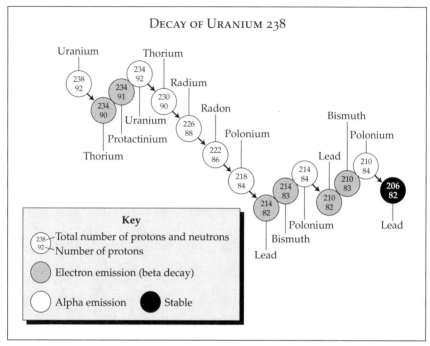

DECAY OF URANIUM 238

Key

238—Total number of protons and neutrons
92—Number of protons

Electron emission (beta decay)

Alpha emission   Stable

Uranium 238 decays naturally, over a predictable period of time, to form lead. Boltwood realized that the accumulation of lead in rocks could be used to determine their age.

Knowing how much helium is produced from each gram of uranium per billion years, Rutherford and his collaborators were able to see that naturally radioactive rocks are often older than 100 million years. The ages of some samples exceeded 500 million years. Moreover, Rutherford was aware that in assigning ages he would have to account for helium that was escaping from the rocks.

Impressed by these results, and trying to eliminate the uncertainties associated with the leakage of helium, Boltwood decided to work on another method of dating. This decision stemmed from his earlier attempts to demonstrate that, as in the case of radium, all uranium minerals contain the same number of atoms of lead per gram of uranium. Chemical data, however, did not confirm this expectation—the measured lead-to-uranium ratios were found to be different in minerals from different locations.

According to Rutherford and Soddy's theory, a spontaneous transformation of uranium into a final product proceeds through a set of steps, in which alpha particles and electrons are emitted, one after another. Boltwood realized that lead must be the final product and that its accumulation could be used to date minerals. By focusing on lead rather than helium, he hoped to reduce the uncertainties associated with the leakage.

Lead, he argued, is less likely to escape from rocks than helium because, once trapped, lead becomes part of a solid structure.

Motivated by these ideas, Boltwood proceeded with the development of the uranium-lead method of dating. To accomplish this, he had to determine the rate at which lead is produced from uranium. Having achieved this, Boltwood started his investigations in 1905, and before the end of the year, he had analyzed twenty-six samples. One of them was identified as 570 million years old. In a formal publication, which appeared in 1907, he described forty-six minerals collected in different locations; their reported ages were between 410 and 2,200 million years old.

Similar results had been reported earlier by Rutherford from his laboratory in Montreal and by the English physicist Robert John Strutt, Lord Rayleigh, from the Imperial College in London, both of whom had used helium methods. Although there was a wide variation in dates, it became clear that many rocks were at least ten times older than what had been calculated by Lord Kelvin.

IMPACT

The main results of the pioneering work of Boltwood and his successors was the realization that geological times must be expressed in hundreds and thousands of millions of years, rather than in tens of millions of years, as advocated by Lord Kelvin. This was particularly significant for the acceptance of the theory of evolution by biologists and, in general, for a better understanding of many long-term processes on Earth.

Geochronology, for example, has been used in investigations of reversals of the terrestrial magnetic field. Such reversals occurred many times during the geological history of Earth. They were discovered and studied by dating pieces of lava, naturally magnetized during solidification. The most recent reversal took place approximately 700,000 years ago.

It is clear, in retrospect, that the discovery of radioactivity affected geochronology in two ways: by providing tools for radiometric dating and by invalidating the thermodynamic calculations of Lord Kelvin. These calculations were based on the assumption that the geothermal energy lost by Earth is not replenished. The existence of radioactive heating, discovered in 1903 in France, contradicted that assumption and prepared scientists for the acceptance of Boltwood's findings. Lord Kelvin died in the same year in which these findings were published, but he knew about Rutherford's findings as early as spring, 1904. He was very interested in radioactive heating but never came forth with a public retraction of his earlier pronouncements.

See also Fossils; Geologic Change; Geomagnetic Reversals; Mass Extinctions; Microfossils; Radioactivity.

FURTHER READING

Asimov, Isaac. *Exploring the Earth and the Cosmos.* New York: Crown, 1982.
Badash, Lawrence. "Bertram Borden Boltwood." In *Dictionary of Scientific Biography*, edited by Charles Coulston Gillispie. New York: Charles Scribner's Sons, 1970.
_____. *Radioactivity in America.* Baltimore: The Johns Hopkins University Press, 1979.
_____, ed. *Rutherford and Boltwood: Letters on Radioactivity.* New Haven, Conn.: Yale University Press, 1969.
Burchfield, Joe D. *Lord Kelvin and the Age of the Earth.* New York: Science History Publications, 1975.
Eicher, Don L. *Geologic Time.* 2d ed. Englewood Cliffs, N.J.: Prentice-Hall, 1976.
Faul, Henry. *Ages of Rocks, Planets, and Stars.* New York: McGraw-Hill, 1966.
Friedlander, Gerhart, et al. "Nuclear Processes in Geology and Astrophysics." *Nuclear and Radiochemistry.* 3d ed. New York: John Wiley & Sons, 1981.

—Ludwik Kowalski

RECOMBINANT DNA TECHNOLOGY

THE SCIENCE: Molecular geneticists pioneered techniques that allow scientists to insert DNA from any source into bacteria and detect the expression of the foreign genes in these simple cells.

THE SCIENTISTS:
Stanley Norman Cohen (b. 1935), American molecular geneticist
Herbert Wayne Boyer (b. 1936), American biochemist
Paul Berg (b. 1926), American biochemist
Hugh Oliver Smith (b. 1929), American molecular geneticist

BACTERIAL HOSTS

Recombinant DNA (deoxyribonucleic acid) technology—known also in various guises as "genetic engineering," "genetic modification," and "gene cloning"—is an area of scientific investigation and applied biology that has, since its inception in 1973, revolutionized molecular biology, allowing scientists to address questions in cell biology that could not be addressed by earlier methods. Recombinant DNA methods allow molecular biologists to add one or a small number of genes from essentially any organism to simple bacterial cells. These foreign genes can be made to become an integral part of the bacterium, replicating along with the bacterial genetic material and thus stably transmitted from one bacterial generation to the next. The foreign genes can also be made to be functional in their bacterial host—that is, they can be induced to make their normal gene products.

Bacteria are very simple single-celled organisms that are ubiquitous in nature. Although some are capable of causing disease, most bacteria are harmless to humans. Some, like the common intestinal bacterium *Escherichia coli* (*E. coli*), are normal inhabitants of the human body that are essential to human life. Each *E. coli* cell has a single circular DNA molecule, or chromosome, containing between two thousand and three thousand genes. In addition, some cells have one or more additional small circular DNA molecules called plasmids. A typical plasmid contains on the order of five to ten genes and is therefore much smaller than the *E. coli* chromosome. These plasmids are semiautonomous, meaning that while they are incapable of leading a cell-free existence, they generally remain separate from the larger chromosome and control and direct their own replication and transmission to each daughter cell at cell division. Plasmids that contain genes for resistance to certain antibiotics, viruses, and so forth can provide the host cell with useful properties.

THE PROCESS

The "basic experiment" of recombinant DNA technology involves four essential elements: a method of generating pieces of DNA from different sources and splicing them back together; a "vector" molecule (often a plasmid) that can replicate both itself and any foreign DNA linked to it; a way to get this composite, or recombinant, DNA molecule back into a suitable bacterial host; and a means to separate those bacterial cells that have picked up the desired recombinant plasmid from those cells that have not.

As part of the process, the recombinant plasmids are then reintroduced back into *E. coli* host cells in a process called "transformation." An essential feature of transformation is treatment of the host cells with calcium chloride, which weakens the cell walls and membranes, allowing the reconstituted plasmid DNA to be taken up inside the cells. If all has gone well,

STANLEY COHEN AND EGF

Stanley Cohen was born November 17, 1922, in Brooklyn, New York, to Jewish emigrants from Russia. He majored in biology and chemistry at tuition-free Brooklyn College, the only college at which he could afford to enroll. Cohen went on to earn his master's degree with a concentration in zoology from Oberlin College in 1945. At the University of Michigan, he earned his Ph.D. in biochemistry in 1948. Then he moved to the University of Colorado, where he studied the metabolism of premature infants.

In 1952, Cohen moved to Washington University in St. Louis and learned the isotope methodology for studying metabolism as a postdoctoral fellow of the American Cancer Society, later working on nerve growth factor with Rita Levi-Montalcini. In 1959, he moved to Vander-

(The Nobel Foundation)

bilt University, where he studied epidermal growth factor (EGF), a protein, which he identified in the early 1960's. Because of the difficulty of amino acid sequencing at that time and the unusual structure of the protein, Cohen and his colleagues were unable to determine EGF's amino acid sequence until the early 1970's. EGF has been found in many different species, including humans (as urogastrone), and is recognized as significant in embryonic and fetal development. The potential of EGF for accelerating wound growth, healing, and growth of skin cells in culture for burn victims prompted the interest of pharmaceutical companies.

Cohen's persistent work on growth factors, at a time when growth factors were not popular in scientific circles and in many cases were held to be suspect, laid the groundwork for another extremely important field that was to develop only in the 1980's with the development of recombinant DNA technology. The ability to clone selected genes from the genetic material (DNA) of the cell and to read the DNA base sequence to determine what protein the gene would make led to the identification of oncogenes, which are viral genes that cause normal cells to develop as tumors. Many oncogenes are now recognized for producing growth factors or proteins that mimic the receptor protein; this protein, however, is always turned "on" to cause cell division. Cohen's work was therefore instrumental in the understanding of mechanisms of cancer induction by viruses.

these genetically engineered clones of bacterial cells will then stably replicate the foreign DNA, along with the rest of the chromosomal and plasmid DNA of each cell generation; the products of the foreign genes—ribonucleic acid (RNA) or protein—will be made as well.

THE FIRST RECOMBINATION

By the early 1970's, the stage was set for the advent of recombinant DNA technology. DNA "ligases" (enzymes that play a significant role in the process) had been discovered and purified independently in five separate laboratories in 1967. Hugh Oliver Smith described the first restriction endonuclease enzyme in 1970, and shortly thereafter Herbert Wayne Boyer described the isolation of EcoRI, a restriction endonuclease that became extremely important in the development of cloning methods. Paul Berg and his group described the construction of the first recombinant DNA molecules in a test tube, and at about the same time researchers in Stanley Norman Cohen's laboratory reported on the first successful transformation experiments in *E. coli.*

In the fall of 1973, Cohen and Boyer were the first researchers to describe successfully a complete recombinant DNA experiment. Their report detailed the mixing and subsequent reconstitution of DNAs from two separate plasmids in *E. coli.* Shortly thereafter, they described experiments in which DNA from a plasmid found in an unrelated bacterium was successfully cloned in *E. coli,* and one year later they reported on the first successful cloning of animal genes in *E. coli.*

IMPACT

Recombinant DNA technology is widely considered to be the most significant advance in molecular biology since the elucidation of the molecular structure of DNA in 1953 by biophysicists James Watson and Francis Crick. It soon became apparent, however, that the technology had opened a Pandora's box of social, ethical, and political issues unprecedented in scientific history. The research held the potential of addressing biological problems of fundamental theoretical and practical importance, yet it generated real concerns also, because some experiments might present new and unacceptable dangers. Even in the course of scholarly research with the best intentions, there was concern that a laboratory accident or an unanticipated experimental result might introduce dangerous genes into the environment, with *E. coli* carrying them.

Soon after the scientific concerns were first voiced, a conference was planned to allow many of the leading researchers in molecular biology to try to assess the potential dangers of recombinant DNA technology. The conference was held at the Asilomar Conference Center in February of 1975. Six months earlier, however, eleven respected authorities in molecular biology, including Cohen, Boyer, Berg, and others who helped develop recombinant DNA techniques, signed a letter that was simultaneously published in three English and American scientific journals. This letter called for a voluntary moratorium on recombinant DNA experiments until questions about potential hazards could be resolved. The development of a set of guidelines for recombinant DNA research, a modification of which was later adopted by the National Institutes of Health, was discussed at the Asilomar Conference. Levels of both biological and physical "containment" were defined, and each type of recombinant DNA experiment was assigned to an appropriate level. Some types of experiments were banned. In the years that followed the initial furor, guidelines have been modified accordingly, as many of the initial fears about possible dangers have proved to be groundless.

As predicted, recombinant DNA technology has proved to have extensive practical applications, particularly in the fields of medicine and agriculture. Virtually all insulin-dependent diabetics now take human insulin made by genetically engineered bacteria. Human growth hormone, prolactin, interferon, and other human gene products with specific therapeutic uses in medicine are available only because they can be made in quantity by using cloning. In agriculture, improved species of genetically modified crop plants have been designed to help address problems in global food supplies. Of particular note is the effort to clone the bacterial genes for nitrogen fixation into crop plants, thus obviating the need for most fertilizers.

See also Chromosomes; Cloning; DNA Fingerprinting; DNA Sequencing; Double-Helix Model of DNA; Evolution; Gene-Chromosome Theory; Genetic Code; Human Evolution; Human Genome; Mendelian Genetics; Mitosis; Oncogenes; Population Genetics; Ribozymes; Stem Cells; Viruses.

Further Reading

Cohen, Stanley N. "The Manipulation of Genes." *Scientific American* 233 (July, 1975): 24-33.
Grobstein, Clifford. "The Recombinant-DNA Debate." *Scientific American* 237 (July, 1977): 22-33.

Jackson, David A., and Stephen R. Stich, eds. *The Recombinant-DNA Debate.* Englewood Cliffs, N.J.: Prentice-Hall, 1979.

Knowles, Richard V. *Genetics, Society, and Decisions.* Columbus, Ohio: Charles E. Merrill, 1985.

Vigue, Charles L., and William G. Stanziale. "Recombinant DNA: History of the Controversy." *American Biology Teacher* 41 (November, 1979): 480-491.

—*Jeffrey A. Knight*

RELATIVITY

THE SCIENCE: Albert Einstein's articulation of special and general relativity not only explained gravitation and motion between contained systems but also created a new view of space and time that laid the foundation for models that have been proposed to explain the creation and evolution of the universe.

THE SCIENTISTS:
Albert Einstein (1879-1955), German-Swiss American physicist
David Hilbert (1862-1943), German mathematician
Sir Arthur Stanley Eddington (1882-1944), English astronomer, philosopher, and physicist

A MATTER OF SOME GRAVITY

Although Sir Isaac Newton's law of gravity was very successful at making predictions, it did not explain how gravity worked. In fact, Newton stated that explaining gravity was not his goal: "Gravity must be caused by an agent acting constantly according to certain laws; but whether this agent should be material or immaterial, I have left to the consideration of my readers." For almost two hundred years, however, Newton's readers cared little about this question. Newton's law can be described as an "as if" law. Two bodies act as if there is a force between them that acts like the force of gravity that Newton proposed. Newton did not address the question of how or why this force operated.

Furthermore, Newton's physics could not explain why, in calculating gravitational attraction between objects, gravitational mass turns out to be exactly equal to inertial mass. (Gravitational mass determines the strength of the gravity acting on an object, and inertial mass determines that object's resistance to any force.) It is because of this equality that all bodies that fall under the influence of gravity alone have the same acceleration, their different masses notwithstanding. It was this strange equality that directed Einstein's thoughts toward a theory of gravity when almost no one else considered it to be an important question.

It's All Relative

When Einstein developed his special theory of relativity in 1905, he did it in an environment in which many other scientists were working on the same problem. The questions involved were the burning issues of the time, and other scientists were also coming close to the answers. The situation in which the general theory of relativity was developed was much different. Einstein did receive some help from his friend Marcel Grossman, a mathematician, and there was a later parallel effort to develop the same theory by the famous mathematician David Hilbert, who was inspired to do so by Einstein. Aside from these minor exceptions, however, Einstein's work on general relativity was entirely his own, and it was performed in an atmosphere in which there was little independent interest in the problem.

Einstein began to work on general relativity after he examined the defi-

Albert Einstein receiving his certificate of U.S. citizenship in 1940. (Library of Congress)

EINSTEIN ON THE SCIENTIFIC IMPULSE

In 1950—more than three decades after he first presented his general theory of relativity—Albert Einstein philosophized on the human impulse toward theoretical science:

What, then, impels us to devise theory after theory? Why do we devise theories at all? The answer to the latter question is simply: Because we enjoy "comprehending," *i.e.,* reducing phenomena by the process of logic to something already known or (apparently) evident. New theories are first of all necessary when we encounter new facts which cannot be "explained" by existing theories. But this motivation for setting up new theories is, so to speak, trivial, imposed from without. There is another, more subtle motive of no less importance. This is the striving toward unification and simplification of the premises of the theory as a whole. . . .

There exists a passion for comprehension, just as there exists a passion for music. That passion is rather common in children, but gets lost in most people later on. Without this passion, there would be neither mathematics nor natural science. Time and again the passion for understanding has led to the illusion that man is able to comprehend the objective world rationally, by pure thought, without any empirical foundations—in short, by metaphysics. I believe that every true theorist is a kind of tamed metaphysicist, no matter how pure a "positivist" he may fancy himself. The metaphysicist believes that the logically simple is also the real. The tamed metaphysicist believes that not all that is logically simple is embodied in experienced reality, but that the totality of all sensory experience can be "comprehended" on the basis of a conceptual system built on premises of great simplicity. The skeptic will say that this is a "miracle creed." Admittedly so, but it is a miracle creed which has been borne out to an amazing extent by the development of science.

Source: Albert Einstein, "On the Generalized Theory of Gravitation: An Account of the Newly Published Extension of the General Theory of Relativity Against Its Historical and Philosophical Background." *Scientific American* 182, no. 4 (April, 1950).

ciencies of the special theory of relativity. Gravity and relativity were incompatible. In particular, in order to incorporate an understanding of gravity into the special theory of relativity, it was necessary to deny the equality of gravitational mass and inertial mass. Because that equality could be established experimentally to a high degree of accuracy, however, it was impossible to ignore. Furthermore, the triumph of special relativity was that it established that all motion was relative. There was no longer any concept of absolute velocity; only the idea of relative velocity remained. Acceleration, however, was left as an absolute. Einstein thought that acceleration should be relative if velocity was relative. The apparent discrepancy, along with the problem of gravity, disturbed Einstein and led him to develop his theory of gravity, which is known as the general theory of relativity.

The crucial step in the development of the general theory of relativity was the publication of the "Principle of Equivalence" in 1907. In this paper, it was proposed that, in any small region of space, one could not distinguish between gravity and acceleration. This meant that a person in a closed room who saw objects fall when they were dropped had no way of knowing whether those objects fell because the room was at rest on the surface of a planet or because the room was in a rocket ship that was accelerating in the direction opposite to the direction of the falling objects. Because gravity and acceleration were equivalent, the equivalence of gravitational mass and inertial mass was thus explained.

In 1911, Einstein used the principle of equivalence to establish that, because light seen by an accelerated observer is bent, gravity must also bend light. Between 1911 and 1916, Einstein worked on developing the complicated mathematics of his complete theory. In doing so, he discovered the correct value for the bending of light. Light follows the shortest distance between two points. When the shortest distance between two points appears to be curved, it means that the area of space that is involved is curved. For example, a straight line drawn on the two-dimensional surface of a globe will, if it is projected onto a flat map, appear to be curved. By using this line of reasoning, Einstein concluded that the curved path of light near a mass means that the four-dimensional space-time around that mass is curved. Furthermore, in 1917, Einstein used general relativity to show that the total mass of the universe affects the structure or shape of the universe as a whole.

IMPACT

Newton's theory of gravity was almost—but not quite—perfect, but Einstein's theory, as Einstein himself found in 1915, corrected those imperfections. More important to the acceptance of this theory, however, was the verification of its predictions for the bending of light. The experiment that was needed required, among other things, a total eclipse of the Sun so that light from the stars could be checked to see whether it was bent as it passed the Sun. This experiment was carried out in Africa in 1919 by the British astronomer Sir Arthur Eddington, and its results verified Einstein's theory.

Since that time, the "geometrification" of space has led to the idea of microscopic "wormholes" connecting one point in space to another. On a larger scale, this geometrification is manifested in searches for "black holes" from which not even light can escape. Such black holes are caused by the extreme warping of space that results when stars collapse. On the largest scale, this new view of space and time provides the basis for the various models that have been proposed to explain the creation and evolution of the universe.

See also Black Holes; Compton Effect; Electron Tunneling; Gravitation: Einstein; Mössbauer Effect; Speed of Light.

Further Reading

Calder, Nigel. *Einstein's Universe: Relativity Made Plain*. New York: Viking, 1979.

Einstein, Albert. *Out of My Later Years*. New York: Philosophical Library, 1950.

Einstein, Albert, and Leopold Infeld. *The Evolution of Physics*. New York: Simon and Schuster, 1938.

French, A. P., ed. *Einstein: A Centenary Volume*. Cambridge, Mass.: Harvard University Press, 1979.

Pais, Abraham. *"Subtle Is the Lord . . .": The Science and the Life of Albert Einstein*. New York: Oxford University Press, 1982.

—*Carl G. Adler*

REM Sleep

THE SCIENCE: Eugene Aserinsky's discovery of rapid eye movements (REMs) in normal human sleep provided the first objective method of studying neural function and behavioral patterns associated with dreaming.

THE SCIENTISTS:
Eugene Aserinsky (1921-1998), graduate student of physiology at the University of Chicago
Nathaniel Kleitman (1895-1999), professor of physiology at the University of Chicago
William Dement (b. 1928), American physiologist

Laboratory Dreaming

As early as 1867, German psychiatrist Wilhelm Griesinger speculated on the occurrence of eye movements during dreams. These eye movements, he believed, occurred both during the transition from wakefulness to sleep and during dreaming. From these observations, he concluded that sleep was not a passive but rather an active state. It was another eighty-five years before Eugene Aserinsky discovered that sleep is not a homogeneous process but is organized in rhythmic cycles of different stages.

In 1952, Aserinsky, a graduate student working on his dissertation in the physiology laboratory of Nathaniel Kleitman at the University of Chicago, achieved a breakthrough in modern sleep research. Aserinsky turned his focus to the study of attention in children, using his young son, Armond, as one of his subjects. While making clinical observations of his young subject's efforts to pay attention, he noticed that eye closure was associated with attention lapse, and thus decided to record these eyelid movements using the electrooculogram (EOG).

Aserinsky and Kleitman observed that a series of bursts of rapid eye movements (REMs) occurred about four to six times during the night. The first such REM period took place about an hour after the onset of sleep and lasted from five to ten minutes. Succeeding REM periods occurred at intervals of about ninety minutes each and lasted progressively longer; the final period occupied approximately thirty minutes.

Suspecting a correlation of eye movements with dreaming, Aserinsky and Kleitman awakened subjects during REM periods and asked them whether they had been dreaming. In a large majority of such awakenings, the subjects acknowledged that they had been dreaming and proceeded to relate their dreams. When subjects were awakened while their eyes were motionless, they could rarely remember a dream. Therefore, Aserinsky and Kleitman concluded that rapid eye movements were an objective signal of dreaming. Although investigators still had to rely upon the dreamer's verbal report to ascertain the content of the dream, the process of dreaming was now opened up to objective study under laboratory conditions.

The Sleep of Cats and Children

In order to obtain a more complete picture of the mental state of his subjects, Aserinsky also recorded brain-wave activity with an electroencephalograph (EEG). Using both the EEG and the EOG enabled Aserinsky to register brain-wave activity during sleep from the moment it began, regardless of the time of day. This combination proved fortuitous because, unlike adults, children often enter the REM phase immediately at sleep on-set, and such sleep-onset REM periods are especially likely to occur during daytime naps.

When Aserinsky's subjects lost attentional focus and fell asleep, their EEGs showed an activation pattern, and their EOGs showed rapid eye movements. Kleitman quickly deduced that this brain-activated sleep state, with its rapid eye movements, might be associated with dreaming. The two investigators immediately applied the combined EEG and EOG measures to the sleep of adult humans and were able to observe the peri-odic alternations of REM and non-REM sleep throughout the night. In ad-dition, when the investigators awakened their subjects during REM sleep, these subjects related accounts of dreams.

In 1953, Aserinsky and Kleitman reported their findings in the journal *Science* in an article titled "Regularly Occurring Periods of Eye Motility and Concomitant Phenomena During Sleep." As is the case with many break-through articles, this one was relatively brief (barely two pages). Yet it in-cluded the observation that during REM sleep, other physiological func-tions vary according to the state of the brain: Respiratory frequency and heart rate fluctuate, and their rhythm becomes irregular.

Physiologist William Dement later confirmed Aserinsky and Kleit-man's hypothesis. Following Aserinsky and Kleitman's groundbreaking 1953 article and their recognition of REM as the physiological basis of dreaming, Dement established that an identical phase of sleep occurs in cats; he published his results in the 1958 *EEG Journal*.

IMPACT

Studies that have attempted to show a relationship between the subject matter of dreams and the physiological changes that occur during REM pe-riods have not established any close correlation between the two phenom-ena. Although early investigations indicated that the pattern of eye move-ments is correlated with the directions in which the dreamer is looking in the dream, subsequent evidence raised doubts concerning this hypothesis.

More conclusive evidence exists to support the theory that dreaming can sometimes occur during non-REM periods. This possibility suggests that dreaming may be more or less continuous during sleep but that condi-tions for the recall of dreams are most favorable following REM awaken-ings. In any case, the prevailing view is that REMs are not an objective sign of all dreaming but that they do indicate when a dream is most likely to be recalled.

Aserinsky's discovery of a stage of sleep during which most dreaming seems to occur led to experiments to investigate what would happen if a sleeping person were deprived of REM sleep. These studies concluded that there is an overwhelming demand for REM sleep. Because REM sleep usually accompanies dreaming, it was also concluded from these studies that there is a strong need to dream. Later studies with prolonged deprivation of REM sleep, however, did not confirm the degree of behavioral changes noted earlier. Thus, it can be concluded that there is definitely a need for REM sleep, but the question of whether there is also a need to dream is still open to debate.

By using an EEG to monitor sleep during the night and by awakening subjects during REM periods, it has been conclusively established that everyone normally dreams every night. Even a person who has never remembered a dream in his or her life will typically do so if awakened during a REM period.

See also Manic Depression; Pavlovian Reinforcement; Psychoanalysis; Split-Brain Experiments.

FURTHER READING

Cohen, David B. *Sleeping and Dreaming*. New York: Pergamon Press, 1979.
Hobson, J. Allan. *The Dreaming Brain*. New York: Basic Books, 1988.
Horne, James. *Why We Sleep*. New York: Oxford University Press, 1988.
Kleitman, Nathaniel. *Sleep and Wakefulness*. Rev. ed. Chicago: University of Chicago Press, 1963.
Oswald, Ian. *Sleeping and Waking*. New York: Elsevier, 1962.
—*Genevieve Slomski*

RIBOZYMES

THE SCIENCE: The demonstration the RNA can act as an enzyme to catalyze biochemical reactions provided evidence of the process of chemical evolution.

THE SCIENTISTS:

Thomas R. Cech (b. 1947), biologist who discovered catalytic RNA
Arthur J. Zaug, colleague of Cech
Sidney Altman (b. 1939), molecular biologist who also discovered the catalytic properties of RNA

Tan Inoue, collaborator with Thomas Cech who later showed the
ability of RNA to catalyze biochemical reactions

PRIMORDIAL SOUP

In the 1920's, Aleksandr Ivanovich Oparin and John Burdon Sanderson
Haldane independently proposed that the early Earth atmosphere lacked
oxygen but contained an abundant amount of hydrogen-containing compounds, such as ammonia, methane, water vapor, hydrogen gas, hydrogen
cyanide, carbon monoxide, carbon dioxide, and nitrogen. They both proposed that these gases spontaneously combined in the presence of energy.
There was no lack of energy on the surface of the early Earth because of
volcanic eruptions, lightning, and ultraviolet radiation. Oparin and
Haldane hypothesized that it was in this type of environment—a reducing
atmosphere without oxygen present—that life on Earth began.

Their model was tested in 1953 by Stanley Miller, a graduate student at
the University of Chicago. He built a system of interconnecting tubes and
flasks designed to simulate the primitive Earth atmosphere and primordial ocean. After a week, he analyzed the results and found simple organic
acids and amino acids, the building blocks of proteins. Miller's experiment
paved the way for others. Utilizing different mixtures of gases, later researchers produced virtually all the organic building blocks necessary for
life and found in cells, including nucleotides, sugars, and fatty acids.

Once all the building blocks were formed, the next important step was
to link these simple molecules into long chains, or polymers. An example
of polymerization would be the linking of amino acids to form a long chain
called a protein. Another polymer would be the polynucleotide, a long
chain of single nucleotides. There are two types of nucleotides, deoxyribonucleic acid (DNA) and ribonucleic acid (RNA), which are very similar in
structure. They differ in that DNA contains the pentose sugar deoxyribose,
while RNA contains the pentose sugar ribose. Ribose has a hydroxyl
group, —OH, instead of a hydrogen atom, —H, at the number 2 carbon
atom. DNA also contains the four nucleotide bases adenine (A), guanine
(G), cytosine (C), and thymine (T), while RNA contains the same nucleotide bases as DNA except that uracil (U) replaces thymine.

Many hypotheses suggest the early polymers may have been formed by
different mechanisms. One means by which the organic molecules might
have been concentrated is the process of evaporation. Another possibility
is that clay particles in the soil, with their characteristic charges that attract
and adsorb ions and organic molecules to their surfaces, might have
brought early organic molecules close enough to one another that they

could polymerize into long chains. The adsorbed metal ions also might have provided a site for the formation of polynucleotides. Once the polynucleotides formed, they then could have acted as a templates specifying "complementary sequences" for the formation of new polynucleotides. These complementary sequences would have resulted from the preferential bonding of certain nucleotides to one another (such as adenine with uracil or thymine, and guanine with cytosine). Geneticists have long known that this simple mechanism accounts for the transfer of genetic information from cell to cell and generation to generation.

The process of polymerization of nucleotides is slow and relatively ineffective; it would have been hindered by the conditions found on the primitive Earth. Even the clay and metal ions would have been slow. Presently, enzymes, which are proteins, function to catalyze (speed up the biochemical reactions involved in) the formation of polynucleotides. In the prebiotic solution or "primordial soup" of the early Earth, however, these enzymes would not yet have been present.

ANCIENT ENZYMES

A discovery in 1981 by Thomas Cech and Arthur J. Zaug indicated how the early polynucleotides might have been replicated. RNA was thought to be a simple molecule, but now this appears not to be the case. Research with the ciliated protozoan *Tetrahymena thermophila* showed the existence of an RNA molecule with catalytic activity. Ribosomal RNA (rRNA) is synthesized as large molecules, which is then spliced to the correct size. In *Tetrahymena*, the surprise came when this reaction was found to occur without the presence of proteins to catalyze the reaction. The only requirement for the reaction to occur was magnesium ions and the nucleotide guanosine triphosphate (GTP). This was a surprising result because in 1981 the dogma in science stated that enzymes were proteins and catalyzed all the reactions of a cell.

Cech later showed that the RNA molecule contained the catalytic activity to splice itself. This self-splicing mechanism resembled the activity of an enzyme and Cech coined the term "ribozyme" to describe the RNA enzyme. A ribozyme is distinguished from an enzyme because it works on itself, unlike other enzymes, which work only on other molecules. Later, Cech studied the properties of the ribozyme and found it similar to enzymes in that it accelerated the reaction and was highly specific. In addition, the three-dimensional structure, as in enzymes, was found to be critical in the activity of the ribozyme. Cech's research showed that if the ribozyme was put in a solution that prevented folding, then the ribozyme showed no catalytic activity, similar to any other enzyme.

The mechanism still needed refinement. Knowing that the folding was important in the catalytic activity aided in discovering the process. Cech and colleagues Brenda L. Bass, Francis X. Sullivan, Tan Inoue, and Michael D. Been discovered that the folding was essential in creating binding sites for GTP. This also activated the phosphate group and increased the likelihood for splitting the RNA molecules. The reaction catalyzed by the ribozyme was speeded up by a factor of 10 billion. Thus, the ribozyme was established as having many enzyme-like properties, such as

Thomas R. Cech. (The Nobel Foundation)

accelerating the reaction and having a three-dimensional structure like an enzyme. It was observed that the ribozyme kept acting on itself. A true catalyst is not converted in the reaction, and the ribozyme was altered in the reaction. In 1983, this distinction also appeared. Zaug and Cech began working with *Tetrahymena* so that a shortened form of the RNA intron could work as a true enzyme.

RNA "Genes"

Another surprise came out of this research. As the ribozyme acted as an enzyme by splicing another RNA chain, it also was synthesizing a nucleotide polymer of cytosine. Not only was the ribozyme acting as a splicer—it was also behaving as a polymerase enzyme by synthesizing chains of RNA molecules that were up to thirty nucleotides long. Later, other researchers strung together strings of nucleotides up to forty-five nucleotides long. These results led to the implication that RNA can duplicate RNA genes.

In 1982, sequences of RNA molecules, introns (intervening sequences in genes), were found to be similar in different types of cells, such as fungi (yeast and *Neurospora crassa*) and protozoans. Remarkably, this discovery of a self-splicing RNA molecule was also found in a bacterial virus. This was a startling discovery, because fungi, protozoans, and viruses were thought to be only very distantly related. A conserved sequence implies an essential function even in the face of evolutionary divergence. This indicates the ribozyme may have evolved relatively early in the evolution of life.

IMPACT

Thomas Cech's work led to geneticists' current understanding of how RNA can duplicate RNA. These conclusions had a profound impact on the theory of the origin of life and chemical evolution. The fact that RNA can act as a catalyst supported the now widely accepted theory of an "RNA world" in which RNA was the primordial genetic material. These functions have now been taken over by DNA and proteins.

It is now known that RNA and DNA store genetic information, but only RNA, as shown by Cech, can act as a catalyst to speed up chemical reactions. Cech's discovery of splicing of RNA by RNA implies that proteins that may have been in existence might not have been needed for gene duplication. The self-splicing of RNA can be considered a primitive form of genetic recombination, since new combinations of RNA sequences are thereby created. Thus, the first genes are thought to be composed of RNA. RNA genes that were combined in a molecule that provided useful products could be at an advantage in the primordial mix.

Cech's discovery that RNA can replicate itself also led researchers to speculate that RNA might catalyze other reactions. Although RNA does not exhibit a high rate of catalytic activity, even a modest rate and some specificity would have been faster than what would have occurred with no enzyme at all.

It is therefore believed that RNA had a significant role to play in the the evolution of life beyond self-replication. If primitive cells, which were surrounded by a membrane, contained these ribozymes, they would have been at a selective advantage over other such cells. Thus, the primitive genetic material of these cells would have been duplicated and passed to other cells. In addition, ribozymes could also bind amino acids in close proximity to allow the amino acids to combine into short polypeptides. These polypeptides could then act as a primitive enzyme, and if they aided the cell in replication and survival of RNA, then the cell could split and pass the genes on to other cells.

Cech's work established a plausible scenario in which RNA might have been the primordial genetic material and enzyme. These functions have been taken over by DNA and proteins, but they are linked together by RNA. The specifics of how life started still remain a mystery, but the pieces

SIDNEY ALTMAN AND RNASE P

In 1983, Canadian biologist Sidney Altman discovered the catalyst RNase P, which consists of both protein and RNA and demonstrated that the RNA is the catalytic part of the molecule. He and his colleagues performed the research that led to this discovery independently of Thomas R. Cech and his team, and virtually simultaneously. For this work Altman shared the 1989 Nobel Prize in Chemistry with Cech.

In his Nobel lecture, Altman gave his perspective on the flow of genetic information within cells, highlighting the role of transfer RNA (tRNA) in the translation of information from RNA into protein. In studying how tRNA is formed from its longer precursor RNA, Altman discovered the enzyme RNase P. This enzyme made a single cut within the longer precursor, cutting it at the site required to produce one particular end of the mature tRNA—an unusually precise reaction for an RNase. Early in his studies of this enzyme, its strong negative charge made Altman suspect that it might be associated with some type of nucleic acid.

(The Nobel Foundation)

Altman's graduate student Benjamin Stark succeeded in purifying and identifying a high-molecular-weight RNA called M1 RNA as a component of RNase P in 1978. He found that the M1 RNA was required for enzymatic activity. Altman described his own reaction: The involvement of an RNA subunit was enough heresy; neither he nor Stark even suspected that the RNA component of RNase P could in itself be a sufficient catalyst for the reaction. Another associate of Altman, Ryszard Kole, then found that the large M1 RNA and the small associated C5 protein of RNase P could be separated into inactive components and then recombined to recover their catalytic ability. By analogy with the ribosome (a very large RNA-protein complex), they began to consider seriously that M1 RNA was contributing to the active site of the enzyme.

At this point, recombinant DNA techniques enabled the Altman group to prepare large quantities of the M1 RNA and the C5 protein and to characterize their structures in detail. In an experiment designed to reconstitute RNase P from two different bacterial species, using *Escherichia coli* M1 RNA and *Baccilus subtilis* C5 protein, Cecilia Guerrier-Takada made the breakthrough discovery. When she tested the M1 RNA alone, under the conditions recommended for the *Baccilus subtilis* RNase P, the M1 RNA alone catalyzed the reaction. The discovery opened the door to speculation of a primordial "RNA world" before the dawn of life.

of the puzzle are coming together. Cech received the 1989 Nobel Prize in Chemistry for this work; he shared it with Sidney Altman, who independently and nearly simultaneously discovered the catalyst RNase P, which consists of both protein and RNA, and demonstrated that the RNA is the catalytic part of the molecule.

See also Chromosomes; Cloning; DNA Fingerprinting; DNA Sequencing; Double-Helix Model of DNA; Evolution; Gene-Chromosome Theory; Genetic Code; Human Evolution; Human Genome; Mendelian Genetics; Mitosis; Oncogenes; Population Genetics; Recombinant DNA Technology; Stem Cells; Viruses.

FURTHER READING

Cech, Thomas. "Ribozyme Self-Replication." *Nature* 339 (June 15, 1989): 507-508.
De Duve, Christian. "The Beginnings of Life on Earth." *American Scientist* 83 (September/October, 1995).
Gesteland, Raymond F., Thomas R. Cech, and John F. Atkins, eds. *The RNA World: The Nature of Modern RNA Suggests a Prebiotic RNA.* 2d ed. Cold Spring Harbor, N.Y.: Cold Spring Harbor Laboratory Press, 1999.
Hart, Stephen. "RNA's Revising Machinery." *Bioscience* 46 (May, 1996).
Horgan, John. "The World According to RNA." *Scientific American* 189 (January, 1996).

—*Lonnie J. Guralnick*

ROSETTA STONE

THE SCIENCE: The discovery of the Rosetta stone provided the key to decipher hieroglyphics, the ancient Egyptian system of writing, and so revealed the long lost, rich culture and history of the civilization.

THE SCIENTISTS:
Jean François Champollion (1790-1832), linguist and Egyptologist who deciphered hieroglyphics
Thomas Young (1773-1829), physician who worked on deciphering hieroglyphics and made invaluable contributions to the understanding of demotic script

NAPOLEON IN EGYPT

In 1798, Napoleon Bonaparte, general of the French military and a national hero, led a military expedition to Egypt. He had defeated most of the enemies of the French Republic except for Britain. He believed that a successful invasion of Britain could not be accomplished until its trade with India was disrupted. To that end, Napoleon planned to conquer Egypt and use it as a military base. Napoleon also had a personal interest in the country, wishing to secure its wealth, strategic value, and potential for development as a French colony. He decided to take 167 scholars with the troops when he left France in May.

Foreigners had ruled Egypt for centuries. The Persians conquered Egypt in 525 B.C.E., were driven out by 380 B.C.E., and returned by 343 B.C.E. The Greeks, led by Alexander the Great, conquered Egypt in 332 B.C.E. By the time of Julius Caesar (100-44 B.C.E.), Egypt no longer spoke its own language. Greek eventually gave way to Latin; the western influence gave way to Arab and then Islamic domination starting in 640 C.E.; and in Napoleon's time the Ottomans ruled the country. When Napoleon's expedition arrived, Egypt had been under the control of the Ottoman Turks for more than three hundred years.

The scholars whom Napoleon brought to Egypt consisted of specialists from all branches of astronomers, engineers, linguists, painters, draftsmen, poets, musicians, mathematicians, chemists, inventors, naturalists, mineralogists, and geographers. Over a three-year period, these "savants" recorded massive amounts of information and provided valuable drawings and sketches that helped spark a renewed interest in Egypt.

A CHANCE DISCOVERY

On August 22, 1798, Napoleon established the Institut d'Égypte (Egyptian Institute of Arts and Sciences) at Cairo, where the savants conducted research and studied the country's history, industry, and nature. On July 19, 1799, a soldier named d'Hautpoul was working to demolish a ruined wall at Fort Rashid (renamed Fort Julien) when he discovered a dark gray stone slab with inscriptions on one side. He reported the discovery to Lieutenant Pierre François Xavier Bouchard, who then informed his superior, Michel-Ange Lancret.

Lancret recognized one of the three scripts as Greek and another as hieroglyphics. The third script was unknown. Bouchard transported the stone to Cairo so that the savants at the institute could examine it. The savants copied the inscriptions using rubbings, drawings, and casts and sent them to scholars throughout Europe so that they could begin working on translating the hieroglyphics.

The Rosetta stone was a basalt slab weighing three-quarters of a ton and measuring 3 feet, 9 inches long, 2 feet, 4.5 inches wide, and 11 inches thick. The stone was damaged, especially the upper portion with the hieroglyphics. The middle section displayed the unknown language—later identified as demotic script—and the bottom portion was Greek.

IDENTIFYING THE CODE

In ancient Egypt, there were two types of writing: hieroglyphics, used in formal writing, and hieratic script, a cursive form of hieroglyphics (simplified and faster), used for everyday writing. By 650 B.C.E., the hieratic script and language had changed so much that it was called demotic. The last known use of hieroglyphics dated from 394 C.E., at a temple in Upper Egypt. Although hieroglyphics had been used for more than three thousand years, no one had read or understood hieroglyphics for fifteen hundred years, so in Napoleon's time the ancient Egyptian civilization was essentially a mystery—lost even though the glyphs were visible on papyrus scrolls, temples, and monuments. By 250 C.E., Coptic—a mixture of demotic and the Greek used by Christian Egyptians—was common in Egypt and marked the first time that vowels had been introduced. Eventually, the Coptic language was replaced, but because it continued to live in the formal Christian liturgy, scholars still could understand the spoken and written forms.

The Greek text on the Rosetta stone was a decree by the priests of Memphis, dated 196 B.C.E., commemorating Ptolemy V Ephiphanes, who ruled Egypt from 204 to 180 B.C.E. According to the decree, Ptolemy V restored the economy and peace, reduced taxes, and was a just ruler, so statues were to be erected and festivals held in his honor. The most exciting part, however, was the conclusion of the text, which indicated that the decree would be inscribed in holy (hieroglyphic), native (demotic), and Greek languages. This directive made it clear that the other parts of the stone essentially were translations of the Greek portion. Reasonably certain that all three scripts recorded the same information, the savants believed that the secret of reading hieroglyphics would be quickly and easily solved.

THE FIGHT OVER ROSETTA

Napoleon left Egypt in August, 1799, to return to France. He took only a few soldiers and some of the savants back with him, as he needed to travel quickly and did not want to appear to give up the Egyptian military campaign. The campaign had failed once the British cut off the supply line, but Napoleon presented the expedition as a success. With severe economic problems fostering a climate for a governmental coup, Napoleon became part of a triumvirate of consuls governing France. In December, 1804, he declared himself Emperor of France.

The remaining troops and savants in Egypt negotiated with the British to leave in early 1800 but were delayed until late 1801. The British wanted to keep all of the records and collections gathered by the savants, but they eventually relented. The British did take back to Britain some major items, however—including the Rosetta stone. Several of the savants decided to go to Britain in order to retain control over their records and collections that the British claimed. Eventually, twenty volumes titled *Description de l'Égypte* (description of Egypt) were published between 1809 and 1828, based on the information collected by the savants. The work covered the

The Rosetta stone.

monuments, natural history, and the modern country as of 1800, and also included the first comprehensive map of Egypt.

In early 1802, the Rosetta stone arrived in Britain and was taken to the Society of Antiquaries in London, where plaster casts were made for universities and engravings were distributed to academic institutions throughout Europe. The stone itself was housed in the British Museum by the end of 1802.

"THIS DECREE SHALL BE INSCRIBED . . ."

The text of the Rosetta stone is dated March 27, 196 B.C.E. Below is an excerpt of the original, unattributed English translation prepared for the British Museum.

DECREE. . . . WHEREAS KING PTOLEMY, THE EVER-LIVING, THE BELOVED OF PTAH, THE GOD EPIPHANES EUCHARISTOS, the son of King Ptolemy and Queen Arsinoe, the Gods Philopatores, has been a benefactor both to the temple and to those who dwell in them, as well as all those who are his subjects. . . .

WITH PROPITIOUS FORTUNE: It was resolved by the priests of all the temples in the land to increase greatly the existing honours of King PTOLEMY, THE EVER-LIVING, THE BELOVED OF PTAH, THE GOD EPIPHANES EUCHARISTOS . . . to set up in the most prominent place of every temple an image of the EVER-LIVING KING PTOLEMY, THE BELOVED OF PTAH, THE GOD EPIPHANES EUCHARISTOS, which shall be called that of "PTOLEMY, the defender of Egypt," beside which shall stand the principal god of the temple, handing him the scimitar of victory, all of which shall be manufactured in the Egyptian fashion; and that the priests shall pay homage to the images three times a day, and put upon them the sacred garments, and perform the other usual honours such as are given to the other gods in the Egyptian festivals; and to establish . . . a statue and golden shrine in each of the temples, and to set it up in the inner chamber with the other shrines; and in the great festivals in which the shrines are carried in procession the shrine of the GOD EPIPHANES EUCHARISTOS shall be carried in procession with them. And in order that it may be easily distinguishable now and for all time, there shall be set upon the shrine ten gold crowns of the king, to which shall be added a cobra exactly as on all the crowns adorned with cobras. . . .

This decree shall be inscribed on a stela of hard stone in sacred and native and Greek characters and set up in each of the first, second and third rank temples beside the image of the ever-living king.

Source: Excerpted from (London, Trustees of the British Museum, 1981). Available at http://pw1.netcom.com/~qkstart/rosetta.html. Accessed September, 2005.

Deciphering Hieroglyphics

Although scholars across Europe worked on translating hieroglyphics, the most important were Jean-François Champollion of France and Dr. Thomas Young of England. Champollion was in Paris by 1807 at age seventeen and working on a copy of the Rosetta stone inscriptions. He realized that hieroglyphics were not only a type of sign but also a hybrid of two classes of language: phonetic (representing sound) and pictorial or ideological (representing pictures or ideas). Eventually, he understood the relationship between hieroglyphics, hieratic, and demotic script, experiencing a breakthrough on September 14, 1822. He later established that hieroglyphics were based on pictograms, ideograms, and phonetic symbols, as well as signs used in special ways. Champollion became the first person able to read hieroglyphics in more than fifteen hundred years.

Young began his work on hieroglyphics in 1814. He realized that some of the hieroglyphics were pictorial, some indicated plurality, and some expressed numbers. He also determined that demotic script uses letters to spell out foreign sounds and was not entirely alphabetic, as some scholars had believed. Young's work on hieroglyphics was published anonymously as a supplement to *The Encyclopaedia Britannica* in 1819. Although Young's system of deciphering did not work, he was the first scholar to make a serious study of demotic script, and his work was invaluable in that regard.

Impact

The discovery of the Rosetta stone made it possible for the first time to unlock the mystery of Egyptian hieroglyphics. Although the stone was discovered in 1799, it would take twenty-three years before hieroglyphics were translated. Once translated, however, the Rosetta stone launched the modern subdiscipline of Egyptology, the study of Egypt and its past.

Both Champollion and Young came to their conclusions independently of one another, and both contributed greatly to understanding hieroglyphic, hieratic, and demotic scripts. By understanding the ancient writing, they helped reveal the history of Egypt and its people to the world. The tremendous amount of written material that had survived on papyri and monuments could now unlock insights into the ancient Egyptian culture that had not been available for any other ancient civilization. Travel to Egypt and the collection and preservation of the ancient monuments and artifacts became a focus of much archaeology over the next two centuries, yielding remarkable information about the complexity of the civilization.

See also Dead Sea Scrolls; Pompeii; Stonehenge; Troy.

FURTHER READING

Adkins, Lesley, and Roy Adkins. *The Keys of Egypt: The Obsession to Decipher Egyptian Hieroglyphs.* New York: HarperCollins, 2000.

Brier, Bob. "Napoleon in Egypt: The General's Search for Glory Led to the Birth of Egyptology." *Archaeology* (May/June, 1999): 44-53.

Meyerson, Daniel. *The Linguist and the Emperor: Napoleon and Champollion's Quest to Decipher the Rosetta Stone.* New York: Ballantine Books, 2004.

Weissbach, Muriel Mirak. "Jean-François Champollion and the True Story of Egypt." *Twenty-First Century Science and Technology* (Winter, 1999/2000): 26-39.

—*Virginia L. Salmon*

RUSSELL'S PARADOX

THE SCIENCE: The logical paradox discovered by Bertrand Russell challenged the long-accepted belief in the consistency of mathematics.

THE SCIENTISTS:

Bertrand Russell (1872-1970), English philosopher
Gottlob Frege (1848-1925), German logician
David Hilbert (1862-1943), German mathematician
Kurt Gödel (1906-1978), Austrian mathematician
Luitzen E. J. Brouwer (1881-1966), Dutch mathematician

MATHEMATICS LOOKS INWARD

The late nineteenth and the early twentieth centuries were characterized by self-reflection within various intellectual domains. For example, Impressionism and later schools of art investigated the very methods of creating art, focusing on "art for art's sake." Sigmund Freud and others founded the field of psychology, which consists of the human psyche looking at itself. Literature, music, architecture, and science (which paid particular attention to the "scientific method") also turned inward.

Mathematics was no exception to this trend; the methods of mathematics were themselves being scrutinized. For example, Gottlob Frege was intensely investigating mathematical logic, the method of mathematical thinking. Central to Frege's work was the mathematically pervasive concept of the "set"—a collection of objects, real or abstract, that could be

defined by either listing its elements or providing a property
that characterized only those elements. For example, one set might
be defined as {2,4,6} or as those
positive even numbers lower than
8. The property-based mode of
definition must be used to define
infinite sets (such as the set of integers), because infinite sets cannot be listed.

Frege's work was well ad-
vanced when Bertrand Russell encountered a peculiar problem in
his definition of sets by properties. Unable to see the solution,
he wrote to Frege in 1902 to in-
quire about the problem. The
older logician replied with one of
the most gracious responses to
bad news ever written, stating

Bertrand Russell. (The Nobel Foundation)

that he had never noticed the problem and that he could not see a solution
for it. Thus, it was discovered that the foundation of Frege's life work was
seriously flawed.

The Paradox

This problem, now called Russell's paradox, is deceptively simple to
delineate. Because sets are well defined, they may be collected into other
sets. For example, the set A may be defined as consisting of all sets with
more than two elements. Set A would therefore contain the set of planets,
the set of negative numbers, the set of polygons, and so forth. Because
these are three sets collected by A, then set A itself has more than two elements. Therefore, set A is a member of itself. This fact may seem strange,
but the defining property is absolutely unambiguous: "sets with more than
two members." Thus, sets may be elements of themselves.

Russell then considered the set D, which consists of those sets that do
not contain themselves. Then he asked, "Does D contain itself?" If it does, it
is one of those sets that it must not contain. Therefore, D must not contain
itself, but D is also one of those sets that it must contain. D contains itself
only if it does not contain itself!

Paradoxes of ordinary language are well known. Two examples are the Cretan Epimenides' remark that "all Cretans are liars" (the liar paradox) and the sentence "This sentence is false." Both statements are true only if they are false. Russell's observation, however, represented the first time that the specter of paradox had arisen within mathematical thought. The seriousness of Russell's paradox stems from the assumption that mathematics embodied a higher truth and was therefore free from error, consistent, and unambiguous. Russell demonstrated that this assumption was false.

IMPACT

Many mathematicians, philosophers, and computer scientists regarded Russell's paradox as an assault on the very foundations of mathematics. If inconsistency could arise in an area as rigorous as set theory, how could consistency be guaranteed in more common areas of mathematics?

Russell and the philosopher Alfred North Whitehead set out to improve upon Frege's work (the Logicist school of thought). If mathematics could be derived from basic, self-evident axioms, no inconsistency would be possible. Russell and Whitehead's *Principia Mathematica* (1910) led to new uses of logic, but its means of avoiding paradox was too arbitrary for all mathematics, since it states that sets cannot contain both objects and other sets.

The Formalist school of David Hilbert, however, sought to establish the foundations of mathematics in the realm of symbol manipulation. The rules that governed such manipulation would be very simple and precise. Such a "proof-theoretic" or "metamathematical" analysis of proof was expected to confirm the consistency of mathematical systems. Much that was useful in mathematics and computer science came out of this work, but in 1931, the young Kurt Gödel astounded the world of mathematics by proving that the Formalist ideal was unreachable—that most mathematical systems could not be proved, by noncontroversial means, to be fully adequate and consistent.

The Intuitionist school of Luitzen E. J. Brouwer grew out of the work of Leopold Kronecker and therefore was not a response to Russell's paradox, but the Intuitionists believed that their insistence on meaning in mathematics would avoid paradox. Intuitionists limit mathematics to what actually can be "constructed" by the human mind. Therefore, infinite sets are ruled out, as is automatic acceptance of the "law of the excluded middle" (which states, basically, that any statement must be either true or false). In this school, the truth or falsity of any statement must be demonstrated. Both of these objections apply to Russell's paradox: D is an infinite set, and the assumption that "D contains D or it does not" is an application of the law of the excluded middle.

The Intuitionist/Constructivist school has not found wide acceptance, because it is viewed as too restrictive by many mathematicians. It is very important, however, in the field of computer science, in which most results must be demonstrated constructively.

It is possible that, before the discovery of Russell's paradox, an easily understood problem had never caused such a major crisis in a scientific field. Russell's paradox had this effect because it forced mathematicians and philosophers to reexamine traditional assumptions about mathematical truth.

See also Abstract Algebra; Axiom of Choice; Bell Curve; Boolean Logic; Bourbaki Project; Calculus; Chaotic Systems; D'Alembert's Axioms of Motion; Decimals and Negative Numbers; Euclidean Geometry; Fermat's Last Theorem; Fractals; Game Theory; Hilbert's Twenty-Three Problems; Hydrostatics; Incompleteness of Formal Systems; Independence of Continuum Hypothesis; Integral Calculus; Integration Theory; Kepler's Laws of Planetary Motion; Linked Probabilities; Mathematical Logic; Pendulum; Polynomials; Probability Theory; Speed of Light.

FURTHER READING

Russell, Bertrand. *The Autobiography of Bertrand Russell*. 3 vols. Boston: Little, Brown, 1967-1969.
Van Heijenoort, Jean. *From Frege to Gödel: A Source Book in Mathematical Logic, 1879-1931*. Cambridge, Mass.: Harvard University Press, 2002.
—*J. Paul Myers, Jr.*

SATURN'S RINGS

THE SCIENCE: After developing an improved technique to grind lenses to precise shapes, Huygens constructed an improved 50-power telescope that helped him identify the unusual elongation of Saturn as a ring or disk surrounding the planet. Huygens also discovered Titan, Saturn's largest moon.

THE SCIENTISTS:
Christiaan Huygens (1629-1695), Dutch astronomer who identified Saturn's rings

Galileo Galilei (1564-1642), Italian astronomer who first observed
   Saturn's rings but thought they were large moons on both sides of
   the planet
Gian Domenico Cassini (1625-1712), Italian astronomer who believed
   Saturn's rings were a multitude of small particles in orbit around
   the planet

GALILEO

In 1610, Galileo the first to observe Saturn with a telescope. He recorded
that Saturn had an odd appearance, with projections that appeared to be
"handles" at both sides. Galileo, however, did not understand his observations. He thought the handles could be two large moons, one on each side
of the planet, so he described Saturn as a group of three, nearly touching
objects that do not move relative to one another. Two years later, in 1612,
Galileo became even more puzzled when he observed that Saturn's "handles" had disappeared.

Although Saturn's ring system was first observed by Galileo, Dutch
physicist and mathematician Christiaan Huygens is credited with their
discovery because he was the first person to identify the observed elongation of Saturn as the presence of a disk or ring surrounding the planet.

CHRISTIAAN HUYGENS

Huygens had studied law and mathematics at the University of Leiden
from 1645 until 1647, and he published a series of papers on mathematics,
but actually he had trained to be a diplomat. In 1649, Huygens was a member of a diplomatic team that was sent to Denmark, but he was not offered a
permanent position in diplomacy. In 1650, Huygens returned home and
lived on an allowance from his father.

Both Huygens and his brother Constantine were interested in astronomy, but they found that the telescopes then available were too short to
resolve features on the planets. The brothers gained an interest in lens
grinding and telescope construction to improve the quality of their observations, and, around 1654, they developed a new and better way of grinding lenses for telescopes. Their techniques significantly reduced "chro-
matic aberration," an effect that causes simple lenses to focus different
colors of light at different points of the telescope lens. They also introduced
the use of "optical stops," masks along the tube of a telescope that intercept
light reflected from the walls of the tube, keeping reflected light from
reaching the lens and blurring the image.

TITAN DISCOVERED

Using one of his own lenses, Christiaan Huygens built a self-designed 50-power refracting telescope. With this new telescope, in 1655, he discovered Titan, the first and largest moon of Saturn. Later that year, he visited Paris and informed the astronomers there, including Ismaël Boulliau (1605-1694), of his discovery. By this time, Boulliau was a well-recognized astronomer who had published his *Astronomia philolaica* (1645), in which he adopted Johannes Kepler's idea that planets moved in elliptical orbits around the Sun. Huygens's discovery of Titan was near the time of the "ring plane crossing" phenomenon, that is, when Saturn's rings are viewed edge-on from the Earth, making them difficult to see. Thus, Huygens was unable to see the rings when he discovered Titan.

THE RING DEBATE

In February of 1656, the true shape of the Saturn's rings was apparent to Huygens. He recognized that the bulge, which Galileo thought were two moons, actually was a thin, flat disk or ring, which did not touch the planet and was inclined to the ecliptic plane. Huygens reported his conclusions in

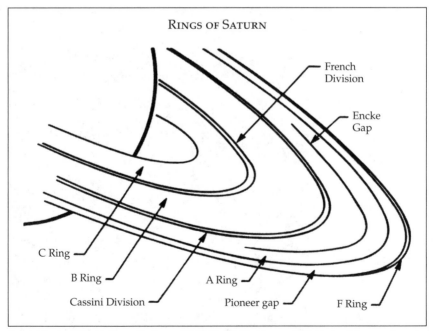

Source: Morrison, David. *Voyages to Saturn*. NASA SP-451. Washington, D.C.: National Aeronautics and Space Administration, 1982, p. 25.

a message to Boulliau, in order to establish the priority of his discovery. However, Huygens did not make a public announcement of his results until 1658, in a letter to the scientific academy in Paris.

Huygens's description of Saturn's rings was not immediately accepted. At least three other astronomers offered different explanations for Saturn's bulge after Huygens's discovery. Gilles Personne de Roberval (1602-1675) proposed that Saturn emitted vapors, like a volcano, from its equatorial region. When the concentration of vapors was high enough, they would become visible as a belt around the planet. Johannes Hevelius, an astronomer from Gdansk, proposed that Saturn was not a sphere, but rather an ellipsoidal, and the bulge was simply part of the planet. Giovanni Battista Odierna (1597-1660) suggested that Saturn had two large dark areas at its equator, which appeared to observers as "handles."

Even with the excellent view of Saturn that Huygens had through his improved telescope, it was not until 1659 that he correctly inferred the geometry of Saturn's rings, because he had to wait until he had observed them over a significant part of their cycle. In his *Systema Saturnium, sive De causis mirandorum Saturni phænomenôn, et comite ejus planeta novo* (1659; the system of Saturn, or on the matter of Saturn's remarkable appearance, and its satellite, the new planet; better known as *Systema Saturnium*), Huygens

Image of Saturn's rings captured by Voyager 2 in 1981.
(NASA/JPL)

explained the phases and changes in the shape of the ring based on the expected view of a rigid disk surrounding the planet and inclined relative to the Earth's orbital path around the Sun. Huygens noted that all earlier observations of Saturn suffered from inadequate resolution. He argued against the models proposed by Roberval, Hevelius, and Hodierna, and offered his idea of a disk surrounding Saturn at its equator but tilted at an angle of about 20° to the plane of Saturn's orbit. He explained that this tilt is what causes the appearance of Saturn's ring to vary as Saturn moves around the Sun.

Although Boulliau generally accepted Huygens's idea of a ring, he believed the ring should still be seen from Earth even when edge-on. Many other astronomers were not convinced. In 1660, Eustachio Divini (1610-1685), an Italian instrument (and telescope) maker, published his "Brevis annotatio in *Systema Saturnium* Christiani Eugenii" (brief comment on Christian Huygens's *Systema Saturnium*), which attacked not only Huygens's ring theory but also the validity of his observations. This book suggested Saturn had four moons, two dark ones near the planet and two bright ones farther out. The handles appeared when the bright moons were in front of the dark ones, partially blocking them from Earth.

Huygens quickly replied with his "Brevis assertio *Systematis Saturnii* sui" (1660; brief defense of *Systema Saturnium*), pointing out that the work of other astronomers contained incorrect observations, which could be explained only by their use of inferior telescopes. Hevelius accepted the ring theory after reading "Brevis assertio *Systematis Saturnii* sui." By 1665, the matter was finally settled, when telescope quality had improved to the point that most astronomers were able to replicate Huygens's observations.

A Solid Ring?

The question that faced the astronomers next was how such a disk could be stable. Huygens thought the ring was a solid structure, but Gian Domenico Cassini proposed that the ring consisted of a large number of small particles, all orbiting around Saturn. Cassini, who conducted extensive observations of Saturn using telescopes at the new Paris Observatory, noted that there was a dark gap separating the ring into two separate rings. This showed that Saturn's rings could not be a single, rigid disk, as proposed by Huygens.

In was not until 1858 that James Clerk Maxwell (1831-1879), a Scottish physicist, was able to perform a detailed mathematical analysis that showed how a ring composed of many tiny particles could be stable. By the end of the nineteenth century, astronomers were able to measure the speed of the particles at the inner and outer edges of the ring. This measurement was inconsistent with a solid rotating disk, and it agreed with the orbital speeds calculated from Kepler's laws of motion.

IMPACT

The rings remained a planetary feature unique to Saturn until 1977, when fainter rings were discovered around Uranus and, shortly after, around the two other gas giant planets, Jupiter and Neptune.

Even more important than these observations, however, was Huygens's insight that the Saturnian system was a miniature solar system, with Titan orbiting Saturn the way Earth orbits the Sun, as Nicolaus Copernicus and Kepler had proposed. Thus, Huygens's observations supported the Copernican idea of a Sun-centered (heliocentric) rather than an Earth-centered (geocentric) solar system. His work was done at a time when a great debate on the issue of a heliocentric versus geocentric system was raging among the best minds in astronomy in Europe.

Because of his great contribution to the understanding of Saturn, the National Aeronautics and Space Administration (NASA) named its Titan space probe the Huygens probe. The probe, which reached Titan in January, 2005, during the Cassini-Huygens mission, fittingly transmitted some of the most important data on Titan to date.

See also Cassini-Huygens Mission; Extrasolar Planets; Galileo Mission; Herschel's Telescope; International Space Station; Jupiter's Great Red Spot; Kepler's Laws of Planetary Motion; Mars Exploration Rovers; Moon Landing; Nebular Hypothesis; Space Shuttle; Voyager Missions.

FURTHER READING

Andriesse, Cornelis D. *Christian Huygens*. Paris: Albin Michel, 2000.

Bell, A. E. *Christian Huygens and the Development of Science in the Seventeenth Century*. London: Edward Arnold, 1947.

Brashear, Ronald. "Christiaan Huygens and His *Systema Saturnium*." http://www.sil.si.edu/DigitalCollections/HST/Huygens/huygens.htm. Accessed February, 2005.

Miedema, Sally. *Huygens: The Man Behind the Principle*. New York: Cambridge University Press, 2005.

Moore, Patrick. *Eyes on the University: The Story of the Telescope*.: New York: Springer-Verlag, 1997.

North, John. *The Norton History of Astronomy and Cosmology*. New York: W. W. Norton, 1995.

Struik, Dirk J. *The Land of Stevin and Huygens: A Sketch of Science and Technology in the Dutch Republic During the Golden Century.* Boston: Kluwer, 1981.

Yoder, Joella G. *Unrolling Time: Huygens and the Mathematization of Nature.* New York: Cambridge University Press, 2004.

—*George J. Flynn*

SCHICK TEST

THE SCIENCE: Béla Schick developed the Schick test, which is performed on the skin to find out how susceptible a person is to diphtheria.

THE SCIENTISTS:

Béla Schick (1877-1967), Hungarian microbiologist and pediatrician

Edwin Klebs (1834-1913) and Friedrich Löffler (1852-1915), German microbiologists who identified the bacteria that cause diphtheria

Pierre-Paul-Émile Roux (1853-1933), French microbiologist

Alexandre Yersin (1863-1943), Swiss microbiologist

Emil von Behring (1854-1917), German microbiologist who discovered a diphtheria antitoxin

A KILLER DISEASE

Diphtheria is a serious disease of the upper respiratory tract—the mouth, nose, and pharynx. The person with this disease may have a fever, a sore throat, and pain all over the body. If the disease is not treated with antibiotics, the infection spreads and causes tissue damage in the heart or kidneys and the victim will eventually die.

Diphtheria is caused by *Corynebacterium diphtheriae*, a rod-shaped species of bacteria. The bacterium can be spread from one person to another by touching or by droplets (for example, from sneezes). Once it enters a person's body, the bacterium releases protein toxins that destroy the membranes and inner structures of cells.

In the 1800's, this disease was not yet understood, but the first steps toward that task were taken. Louis Pasteur, Robert Koch, and other microbiologists (scientists who study organisms too small to be seen by the naked eye) established the germ theory of disease, showing that infectious diseases are carried by microorganisms, usually bacteria or viruses. Diphtheria, typhoid fever, scarlet fever, tuberculosis, and several other diseases were major killers in the nineteenth century, especially among patients in hospitals. Microbiologists were determined to discover the microorganisms that caused these diseases.

In 1883, the German microbiologists Edwin Klebs and Friedrich Löffler raised guinea pigs infected with diphtheria. Under microscopes, they observed rod-shaped bacteria growing in blood samples from the infected animals. When this bacterium was injected into healthy guinea pigs, they became ill with diphtheria, too. In this way, the scientists proved that this rod-shaped bacterium was the cause of diphtheria.

Löffler believed that these bacteria hurt their victims by releasing a toxin (a chemical that damages cells). In 1888, microbiologists Pierre-Paul-Émile Roux and Alexandre Yersin, working together at the Pasteur Institute in Paris, separated the diphtheria bacteria from the serum in which they were being grown. Roux and Yersin then injected the bacteria-free serum into healthy animals. The animals soon came down with diphtheria, even though they had not been exposed to the bacteria. The scientists realized that a toxin was being released by the *Corynebacterium diphtheriae* into the growth serum, and that it was this toxin that made people and animals ill.

Fighting Diphtheria

Now microbiologists could go to work finding a vaccine and designing methods of diagnosing and treating diphtheria victims. In 1890, German microbiologist Emil von Behring discovered that the blood of animals infected with diphtheria produced an antitoxin, a chemical that binds to a toxin and makes it harmless. Behring realized that the antitoxin might be helpful in producing a vaccine to protect people against diphtheria. He injected animals with weakened diphtheria toxin—just enough so that their immune systems would create antitoxin but not enough to hurt the animals. Unfortunately, this diphtheria toxin was too dangerous to use on humans. Behring's work, however, led to the later use of diphtheria antitoxin produced in horses as a treatment for human victims of the disease. In 1923, a formalin-treated toxin was used to vaccinate people against diphtheria and was found to be safe.

Detecting Susceptibility

In 1908, Béla Schick, a pediatrician and microbiologist from Boglár, Hungary, became an assistant to Theodor Escherich at the University of Vienna, Austria. These two scientists began studying diseases caused by bacteria, including diphtheria and scarlet fever.

In 1913, Schick used Behring's work with antitoxins to develop a test that would show how susceptible a person was to catching diphtheria. The result was the Schick test, which proved to be simple and reliable. About 0.1 milliliter of a weakened toxin solution is injected just under the skin inside a patient's arm. The toxin is treated so that it will lead to a bit of swelling in susceptible persons without hurting them. If the patient is susceptible to diphtheria, a reddened, swollen rash (caused by damaged skin cells) will appear around the injection site within a few days. A person who is not susceptible will have no reaction, since the toxin is not causing damage.

Those who test positive with the Schick test should be immunized. People who are already suffering from diphtheria can be treated with a combination of antibiotics and horse serum antitoxin. Antibiotics destroy the *Corynebacterium diphtheriae* bacteria, while the horse serum antitoxin destroys the diphtheria toxin until the victim's body is strong enough to make enough of its own antitoxin.

IMPACT

Schick's test for diphtheria became a valuable tool for identifying the disease and which people most needed immunization. For his findings, Schick was named Extraordinary Professor of Children's Diseases at the University of Vienna in 1918. His test saved thousands of lives, especially among children, who tend to be susceptible to diphtheria. In the middle-to-late 1920's, when the first successful toxoid vaccine was available, the number of cases of diphtheria around the world dropped dramatically.

During the first two decades of the twentieth century, before the test and vaccine were available, there were between 150,000 and 200,000 diphtheria cases every year in the United States alone. By the 1970's, the number of diphtheria cases had dropped to ten per year.

The work of Schick and others also helped show how microorganisms are present everywhere in the environment and can cause disease once they are inside the human body. This led to a better understanding of the importance of antisepsis and sterilization. Before the 1900's, surgical instruments were kept clean, but they were never sterile (clear of all microorganisms); as a result, many patients died after surgery. Microbiological research in Schick's day led to the sterilization of surgical equipment, antiseptic treatment to keep all hospital rooms and equipment clean, and the sanitation of water.

See also Diphtheria Vaccine.

FURTHER READING

Audesirk, Gerald J., and Teresa E. Audesirk. *Biology: Life on Earth.* 2d ed. New York: Macmillan, 1989.

Breed, Robert S., E. G. D. Murray, and Nathan R. Smith, eds. *Bergey's Manual of Determinative Bacteriology.* 7th ed. Baltimore: Williams & Wilkins, 1957.

Eisen, Herman N. *Immunology: An Introduction to Molecular and Cellular Principles of the Immune Responses.* 2d ed. Philadelphia: J. B. Lippincott, 1980.

Gebhardt, Louis P. *Microbiology.* 4th ed. St. Louis: C. V. Mosby, 1970.

Mader, Sylvia S. *Biology.* 3d ed. Dubuque, Iowa: Win. C. Brown, 1990.

Raven, Peter H., and George B. Johnson. *Biology.* 2d ed. St. Louis: Times Mirror/Mosby, 1989.

Wallace, Robert A., Jack L. King, and Gerald P. Sanders. *Biosphere: The Realm of Life.* 2d ed. Glenview, Ill.: Scott, Foresman, 1988.

—*David Wason Hollar, Jr.*

SCHRÖDINGER'S WAVE EQUATION

THE SCIENCE: Erwin Schrödinger proposed that electrons in an atom travel in waves like those of light.

THE SCIENTISTS:

Erwin Schrödinger (1887-1961), Austrian physicist
Louis de Broglie (1892-1987), French physicist
Niels Bohr (1885-1962), Danish physicist
Werner Heisenberg (1901-1976), German physicist

WAVES IN MATTER?

The first three decades of the twentieth century were a time of great change in the manner in which scientists viewed the world. In this short span of time, the early concept of the atom as an indivisible piece of matter was transformed into a picture of an atom made of different particles with different properties interacting with one another to form one unit. In 1922, Niels Bohr was awarded the Nobel Prize in Physics for his theory that the electrons in an atom orbit the nucleus only at certain distances, or energy levels. These levels are determined by a constant discovered by Max

Planck (1858-1947), which relates to units, or quanta, of light. Although Bohr's theory was widely acclaimed by such eminent scientists as Albert Einstein (1879-1955), there were still problems with it.

Louis de Broglie expanded on Bohr's and Planck's work in his hypothesis that, if light could have some of the properties of particles, perhaps matter could have some of the properties of waves. In thinking about the possibility of "matter waves," he was especially interested in finding exact positions of the electrons in an atom by treating them as if they moved in the same wavelike patterns as those of light. In 1923, de Broglie suggested that the energy levels Bohr had found were simply certain numbers multiplied by Planck's constant.

STANDING WAVES

One of de Broglie's biggest problems was that he could not find enough mathematical evidence to support his theory. Bohr had believed that electrons orbited the nucleus in the same way that planets orbit the Sun. De Broglie realized that atomic structure was more complicated than this, and experimental evidence was soon found that backed him up. No means had been found, however, to predict where an electron would go in any particular atom.

At this time, Erwin Schrödinger was teaching at the University of Zurich, Switzerland, in the same position that Einstein had occupied several years before. Schrödinger was fascinated by Bohr's and de Broglie's theories but found several flaws in them. Both of these earlier theories assumed that the light waves that could be associated with certain atoms (their "spectra") came from electromagnetic waves radiating from the atoms. Schrödinger studied these atoms from a slightly different point of view: He believed that the energy levels found by the earlier theories were the result of "standing waves"—that is, waves that overlapped each other so exactly that they did not allow any other radiation to escape—rather than of continuously radiating energy. Radiation could be detected only when electrons moved from one energy level to another, while the overlap of their paths did not form a standing wave.

Schrödinger developed a complicated equation, known as the Schrödinger wave equation, that could be used to predict where in an atom an electron would be at a certain time. He presented this equation, along with its development, support, and consequences, in a series of six papers published in the last half of 1926. It was quickly shown that the values found by calculating Schrödinger's wave equation for certain numbers corresponded exactly to Bohr's energy levels, as well as to other data, such as the spectral lines of the chemical elements (lines of certain colors of light that depend on the wavelength of the light; the chemical elements have their own, discrete spectral emissions).

HEISENBERG'S MATRIX MECHANICS

At almost the same time that Schrödinger was working on his wave-mechanical view of the atom, Werner Heisenberg was using much of the same data to develop a "matrix-mechanical" view of the atom. Heisenberg's model was based entirely on experimental evidence rather than on ideas of what the atom should look like. The two scientists published their discoveries within a year of each other, and Schrödinger soon showed that he could generate the same results with his wave equation that Heisenberg had with his matrices. The combination of these two theories gave a firm basis for the complete theory called "quantum mechanics."

IMPACT

Although Schrödinger received the Nobel Prize in Physics for his work in 1933, he was not completely satisfied with his conclusions. He had not eliminated from the earlier theories the concept of quantum jumps, or electrons "jumping" through space while going from one energy level to another. Although he had given a logical explanation of the reasons for this occurrence, the existence of quantum jumps was one of the original flaws he had found in Bohr's theory. Various aspects of this flaw continued to be bones of contention among Schrödinger, Heisenberg, and other physicists for years to come, although the more established names in theoretical physics, such as Einstein and Planck, accepted Schrödinger's theories enthusiastically.

Schrödinger's wave equation gave rise to a whole new branch

Erwin Schrödinger. (The Nobel Foundation)

of physics, became the basis for virtually all subsequent developments in the field of chemistry, and had a great impact on many other areas of science, including astronomy and biology. In chemistry, Schrödinger's equation has been used to explain bond energies and bond lengths between the

---

SCHRÖDINGER'S CAT

In quantum mechanics, the laws of physics are governed by probability. Unlike the rest of science, quantum mechanics does not offer models of what will happen given a certain set of circumstances; quantum mechanics merely describes how probabilities change with time. Upset by the absurd implications of this position, Erwin Schrödinger in 1935 framed a famous thought experiment designed to expose it:

> One can even set up quite ridiculous cases. A cat is penned up in a steel chamber, along with the following diabolical device (which must be secured against direct interference by the cat): in a Geiger counter there is a tiny bit of radioactive substance, so small that perhaps in the course of one hour one of the atoms decays, but also, with equal probability, perhaps none; if it happens, the counter tube discharges and through a relay releases a hammer which shatters a small flask of hydrocyanic acid. If one has left this entire system to itself for an hour, one would say that the cat still lives if meanwhile no atom has decayed. The first atomic decay would have poisoned it. The Psi function for the entire system would express this by having in it the living and the dead cat (pardon the expression) mixed or smeared out in equal parts.

The experiment is constructed such that the detector is switched on long enough so that there is a fifty-fifty chance that one of the atoms in the radioactive material will decay and that the detector will record the presence of a particle. If the detector does record such an event, the poison container is broken and the cat dies. If the detector does not record the presence of a particle, the cat lives. In the world of ordinary experience, there is a fifty-fifty chance that the cat will be killed. Without examining the contents of the box, it is safe to assert that the cat is *either* dead *or* alive.

However, if one accepts quantum mechanics, then the atomic decay has neither occurred nor not occurred—*until* one opens the box and observes the outcome—and since the fate of the feline is tied to the state of the radioactive material, one cannot assert the simple truth (on a macro level) that the cat must be either dead or alive. This implication, Schrödinger declared, revealed the absurdity of quantum mechanics.

Source: Quotation from Erwin Schrödinger, "Die gegenwartige Situation in der Quantenmechanik." *Naturwissenschaften* 23 (1935). Translated by John D. Trimmer in *Proceedings of the American Philosophical Society* 124 (1980).

atoms in a molecule, and it continues to suggest other properties of chemical bonds. The area of molecular biology developed from the introduction of the theories of quantum mechanics into chemistry. Quantum theory has also made great contributions to astronomy, affecting research in such subjects as the composition of the Sun and stars, the rate at which stars generate energy, and the structure of stars.

Other scientists skirted the edges of quantum theory, but no other provided such concrete evidence as Schrödinger had with his wave equation. Few other discoveries have had such far-reaching implications for the future of science. Scientists are still finding new applications and implications of Schrödinger's work, and they will continue to do so for a long time to come.

See also Exclusion Principle; Gravitation: Einstein; Heisenberg's Uncertainty Principle; Lasers; Photoelectric Effect; Quantum Mechanics; Spectroscopy; Speed of Light; Superconductivity; Superconductivity at High Temperatures; Thermodynamics: First and Second Laws; Thermodynamics: Third Law; Wave-Particle Duality of Light; X Radiation; X-Ray Crystallography; X-Ray Fluorescence.

FURTHER READING

Gribbon, John. *In Search of Schrödinger's Cat: Quantum Physics and Reality.* New York: Bantam Books, 1984.
Schrödinger, Erwin. "Die gegenwartige Situation in der Quantenmechanik." *Naturwissenschaften* 23 (1935): 807ff. English translation in John D. Trimmer, *Proceedings of the American Philosophical Society* 124 (1980): 323-338.
_____. *What Is Life? The Physical Aspect of the Living Cell.* Cambridge, England: Cambridge University Press, 1944.

—*Margaret Hawthorne*

SCIENTIFIC METHOD: ARISTOTLE

THE SCIENCE: Aristotle was the first philosopher to approach the study of nature in a systematic way, establishing science as a discipline and providing a starting place for natural philosophers into the late Middle Ages.

THE SCIENTIST:
Aristotle (384-322 B.C.E.), founder and head of the Lyceum in Athens

ORGANIZING KNOWLEDGE

Born in Stagira in northern Greece and the son of the physician to Amyntas II of Macedonia (r. c. 393-370/369), Aristotle came to Athens when he was seventeen years old and studied at Plato's Academy for twenty years. When Plato died in 347 B.C.E., Aristotle left the academy and traveled for twelve years, visiting various centers of learning in Asia Minor and Macedonia. During this period of travel, he developed his interest in the natural sciences, to which he applied his method of inquiry. He returned to Athens in 335 B.C.E. after a brief period of tutoring Alexander the Great (356-323 B.C.E.), Amyntas's grandson, and established the Lyceum, a school that became a center of learning. He taught there until a year before his death.

The range of topics discussed and developed by Aristotle at the Lyceum is overwhelming: natural philosophy with its considerations of space, time, and motion; the heavenly bodies; life and psychic activities; ethical and political problems; animals and biological matters; and rhetoric and poetics. He is sometimes credited with creating new fields of research, such as terrestrial dynamics and optics. He also taxonomized plants and animals and organized earlier Greeks' ideas about planetary astronomy in *Peri ouranou* (c. 350 B.C.E.; *On the Heavens*, 1939).

Perhaps the most significant aspect of Aristotle's work is his development of a "scientific" approach to these studies. This approach recognizes the existence of independent disciplines, each employing its own principles and hypotheses. Such an approach also works out a methodology or procedure for each field of study, aiming at true and certain knowledge.

The Greek term that Aristotle uses for "scientific knowledge" is *episteme*, which can best be translated as "true knowledge" or the "most certain knowledge." This knowledge includes the awareness of an object, of its causes, and that it can be no other way. Medieval scholars translated the Greek episteme as the Latin scientia, which came into English as "science."

In recognizing independent fields of study, Aristotle showed a significant departure from Plato's philosophy. Plato had envisioned one single science. For him, true knowledge was the contemplation of the Forms: Virtue, Justice, Beauty, and Goodness. All other disciplines were subordinate to knowledge of the Forms. Aristotle, on the other hand, did not advocate a hierarchical structure of knowledge. Each study locates its own particular subject matter and defines its principles from which conclusions are to be

ARISTOTLE'S SCIENTIFIC METHOD

Students learn from a young age that the "scientific method" involves an empirical approach to reality:

(1) Make observations (empiricism).
(2) Develop a tentative explanation, or "hypothesis."
(3) Make predictions of fact based on the hypothesis.
(4) Develop experiments to test the predictions of the hypothesis.
(5) Note where the hypothesis fails and refine it.

Aristotle, by contrast, counseled in his *Posterior Analytics* that science must proceed from "primary premises"—self-evident principles or propositions, based in fact and stating a logically indemonstrable truth:

> [I]t is clear that we must get to know the primary premises by induction; for the method by which even sense-perception implants the universal is inductive. Now of the thinking states by which we grasp truth, some are unfailingly true, others admit of error—opinion, for instance, and calculation, whereas scientific knowing and intuition are always true: further, no other kind of thought except intuition is more accurate than scientific knowledge, whereas primary premises are more knowable than demonstrations, and all scientific knowledge is discursive. From these considerations it follows that there will be no scientific knowledge of the primary premises, and since except intuition nothing can be truer than scientific knowledge, it will be intuition that apprehends the primary premises—a result which also follows from the fact that demonstration cannot be the originative source of demonstration, nor, consequently, scientific knowledge of scientific knowledge. If, therefore, it is the only other kind of true thinking except scientific knowing, intuition will be the originative source of scientific knowledge. And the originative source of science grasps the original basic premise, while science as a whole is similarly related as originative source to the whole body of fact.

What Aristotle called his scientific method was actually somewhat deductive, as Francis Bacon would later point out. Aristotle's "primary premises" arose from sense perceptions, gathered over time and formed intuitively into abstractions—hence, from the many (sense experiences) came the one (premise). To Aristotle, this was induction. However, when sense perceptions are faulty, incomplete, or inaccessible (not all facts can be detected by the human senses alone), the premise would be in error as well, no matter how logically consistent it might be. Still, Aristotle's belief in working from the particular to the general laid the foundation for scientific methodology for nearly two millennia, until Bacon made it truly inductive.

Source: Aristotle, *Posterior Analytics*, translated by G. R. G. Mure. Available at The Internet Classics Archive, http://classics.mit.edu. Accessed September, 2005.

Aristotle. (Library of Congress)

drawn. Almost all his treatises begin with the same format: "Our task here concerns demonstrative science" (that is, logic) or "Human conduct belongs to political science."

Aristotle's insistence on the division of sciences, each using special principles, is indicative of his rejection of any absolute master plan of knowledge. He does, however, recognize "common principles," or principles shared by more than one science. For example, the "equals from equals" principle of mathematics can be used in geometry to deduce a conclusion about a line. Aristotle warns the geometrician, however, that this can be done "if he assumes the truth not universally, but only of magnitudes." Aristotle never intends the same common principles to be universally applied in exactly the same way throughout all the sciences. If this were the case, there would not be "sciences," but rather "Science."

ARISTOTLE'S METHODOLOGY

One of the most important features of Aristotle's scientific approach concerns methodology. In the *Analytica posterioria* (335-323 B.C.E.; *Posterior Analytics*, 1812), he develops the general technique that the particular disciplines are to employ in order to achieve scientific knowledge. First, an investigation must always begin with what is "better known" to humans—with observable data and facts—and not construct wild hypotheses. Second, human beings must proceed to a knowledge of the cause of the facts; mere observation is not enough. Observing something only indicates that something is the case; it does not explain why it is the case. Learning the cause tells people why, and this involves a logical demonstration. Third, the cause or reason of the fact must be of "that fact and no other." This criterion is the basis for a scientific law because it demands a universal connection between the subject and its attributes.

DEDUCTION AND INDUCTION

The second and third criteria require a deductive system of demonstration that is expressed in the universal positive form of the syllogism that Aristotle developed in the *Analytica priora* (335-323 B.C.E.; *Prior Analytics*, 1812). There is also what might be called an "inductive" approach to his method of science. Aristotle raises the question of how humans know the universal principles from which demonstration is to proceed. He answers that human knowledge of such principles begins with many sense perceptions of similar events. Human memory unifies these perceptions into a single experience. The human intellect or mind then understands the universal import of the experience. From many similar experiences, humans recognize a universal pattern.

Aristotle's method of science combines the theoretical and the practical. The theoretical aspect includes logical demonstrations and universal principles. The practical includes the necessary role of sense perception as it relates to particular objects. In the *Metaphysica* (335-323 B.C.E.; *Metaphysics*, 1801), he warns that physicians do not cure men-in-general in a universal sense; rather they cure Socrates or Callias, a particular man. He adds that one who knows medical theory dealing with universals without experience with particulars will fail to effect a cure. Instead, he advises the use of procedures grounded in common sense that have proven their validity in practice.

One application of this method is in Aristotle's writings on biology. He makes theoretical interpretations based on his dissection of marine animals and empirical observations, although he does also rely on other writers' descriptions of some animals. Based on these researches, he arranges a "ladder of nature." Because he can see changes in the realm of plants and animals, he affirms the reality of nature and the value of its study. He is optimistic that he could use natural history to find causal explanations of physiology.

IMPACT

For Aristotle, scientific knowledge included the observation of concrete data, the formulation of universal principles, and the construction of logical proofs. Greek "science" prior to Aristotle, largely a melange of philosophical and quasimythological assumptions, blossomed after his investigations into the specialized work of Theophrastus (c. 372-c. 287 B.C.E.) in botany, Herophilus (c. 335-c. 280 B.C.E.) in medicine, and Aristarchus of Samos (c. 310-c. 230 B.C.E.) in astronomy.

Aristotle also pioneered the notion that there are many, distinct disciplines of knowledge rather than a single, unified science; that there are multiple structuring principles for these disciplines rather than one, overarching set of concepts applicable to them all; that standards of scientific rigor vary among disciplines; and that there is no single, universal scientific method. At the same time, he believed in systematic, empirical investigation of natural phenomena, from which general theories might arise, as opposed to creating a theoretical structure and then fitting the data into it. Although the Aristotelianism that survived in the Middle Ages would be questioned and corrected by natural scientists and philosophers such as Francis Bacon, Aristotle's identification of many of the scientific disciplines and his methodology for studying them remain largely valid today.

See also Galen's Medicine; Greek Astronomy; Greek Medicine; Heliocentric Universe; Herschel's Telescope; Medieval Physics; Scientific Method: Bacon; Scientific Method: Early Empiricism.

FURTHER READING

Barnes, Jonathan, ed. *The Cambridge Companion to Aristotle*. New York: Cambridge University Press, 1995.
Byrne, Patrick H. *Analysis and Science in Aristotle*. Albany: State University of New York Press, 1997.
Ferejohn, Michael. *The Origins of Aristotelian Science*. New Haven, Conn.: Yale University Press, 1991.
Gotthelf, Allan, and James G. Lennox, eds. *Philosophical Issues in Aristotle's Biology*. Cambridge, England: Cambridge University Press, 1987.
Lindberg, David C. *The Beginnings of Western Science*. Chicago: University of Chicago Press, 1992.
Sfendoni-Mentzou, Demetra, et al., eds. *Aristotle and Contemporary Science*. 2 vols. New York: P. Lang, 2000-2001.
                                        —*Joseph J. Romano and Amy Ackerberg-Hastings*

SCIENTIFIC METHOD: BACON

THE SCIENCE: Sir Francis Bacon's *Novum Organum* established an impressive agenda for modern science and inspired the work of later groups, such as the Royal Society of London.

THE SCIENTIST:
Sir Francis Bacon (1561-1626), lord chancellor of England, 1618-1621

SURVEYING THE STATE OF KNOWLEDGE

In the early seventeenth century, the Renaissance was at its height in England and a new age of exploration and scientific instruments had yielded discoveries that required a rethinking of how knowledge was organized, assimilated, and consumed. England was still full of "Renaissance men" with the financial means to avoid narrow specializations and, in Francis Bacon's famous phrase, to "take all knowledge for their province." A number of learned women thrived as well—including Bacon's mother, who translated a religious work from Latin—and the next generation saw the emergence of "female virtuosos" such as the poet and chemist Margaret Cavendish, duchess of Newcastle. Bacon, however, undertook to organize the new learning and to mobilize the students for the monumental task of perfecting God's creation.

The first part of Bacon's great plan was a survey of the arts and sciences. He made his preliminary survey in *The Twoo Bookes of Francis Bacon of the Proficience and Advancement of Learning, Divine and Humane* (1605; enlarged as *De Augmentis Scientiarum*, 1623; best known as *Advancement of Learning*), which he dedicated to the new king of England, James I. In the treatise, he judged certain sciences to have reached a degree of "proficiency," detailed the "deficiency" of others, and made recommendations for their improvement. It was like attempting to write an entire university catalog from scratch.

BACON'S *NOVUM ORGANUM*

In 1620, the second part of Bacon's plan was ready, a description of his method for the brave new world of learning. He called his method the *Novum Organum* (1620; English translation, 1802), or new system of rules, and in doing so, he announced that his method would replace the old rules. It was an outrageously ambitious book. Ever since the rise of the universities in the Middle Ages, the six books of Aristotle's logic had dominated the curriculum. Collectively known as the *Organum* or *Organon*, they were enshrined as the final authority in all debate under the Elizabethan Statutes at Cambridge University, where Bacon had studied. Aristotle's logic was based on syllogism and on deduction from universal precepts to specific conclusions. Bacon's method, by contrast, worked by induction from observations to axioms.

Bacon wrote in Latin so he could reach an international audience. He planned a Latin translation of the *Advancement of Learning* and presented the *Novum Organum* as the second part of a vast work that he called the *Instauratio Magna* (great restoration). He explained that he wanted to help restore human knowledge to the condition that Adam was said to have had in Paradise, before the Fall, and to restore human communication to the universal language that humankind was said to have had at Babel, before the confusion of tongues described in the Old Testament book of Genesis. The large folio volume, printed in London, had an engraved title page that showed a ship sailing beyond the Pillars of Hercules, as the Straits of Gibraltar were once called, and thus going beyond the lands known to the ancient world. The motto on the title page was taken from a prophecy in the Book of Daniel, translated in the King James Bible to read "Many shall run to and fro, and knowledge shall be increased." The implication was that the discoveries in the new age of science, along with the geographical discoveries in the age of exploration, would lead humankind into a golden age.

The *Novum Organum* began with a personal statement, "Francis of Verulam reasoned thus with himself and judged it to be for the interest of the present and future generations that they should be made acquainted with his thoughts." Bacon voiced his fear that the thoughts would die with him

Francis Bacon. (Library of Congress)

if they were not written down and made public. In the preface that followed, he explained that his great work would have six parts.

The first was his survey of learning to date, which had not yet been translated from English. The second part was "The New Organum: Or, Directions Concerning the Interpretation of Nature" and provided the method for what was to follow. The third part would record the "histories" of all the natural and experimental sciences. The fourth would be a set of "instances" discovered about the sciences and pointing to further experiments. The fifth would be a list of axioms that could be inferred provisionally from these instances. The sixth, which would have to be written by Bacon's heirs in a later age, would be the true science toward which he looked. This was a sign of modesty on Bacon's part. His fragmentary notes for part three included 130 subjects for "histories." Here were proposed histories of the elements he knew: fire, air, water, and earth. It would take a later age and the atomic theory to understand hydrogen, oxygen, carbon, and the many other elements.

The *Novum Organum* itself was divided into two books, each of which was written in a series of numbered paragraphs, called "aphorisms" in the translation. The first book discussed the nature of knowledge and the obstacles to knowledge, and among these obstacles included the "idols" that people fashion—distortions that can be tracked to human nature or to individual quirks, to the use of language or the abuse of a philosophical system. The second book was a demonstration of the inductive method he proposed. Here, Bacon set forth an example by investigating the property of heat and creating separate "tables" for studying the presence of heat, the absence of heat, and the increase or decrease of heat. He dedicated his work, once again, to King James. The king wrote a letter of thanks, promising to read the book, but probably never did.

THE RESTORER OF SCIENCE

Two centuries later, Thomas Macauley remarked, famously, that Bacon wrote philosophy like a lord chancellor. Bacon was actually appointed lord chancellor of England in 1618, reaching the peak of the legal profession and marking the end of a long ascent that had taken him from solicitor general to attorney general and lord keeper of the seal. Bacon thought he was in a unique position to dispense justice. At times in his writings, he seems quite highhanded as he presides over the arts and sciences. At times, he is dead wrong. For example, he is sometimes said to underestimate the importance of mathematics.

Bacon was made Lord Verulam in 1618, when he became lord chancel-

---

BACON'S METHOD: INDUCTION VS. DEDUCTION

In Novum Organum, *Francis Bacon addressed the "Interpretation of Nature and the Kingdom of Man" in a series of aphorisms that expressed the importance of drawing inferences inductively from empirical evidence, rather than deductively, from logic, as Aristotle had advocated:*

XII. The logic now in use serves rather to fix and give stability to the errors which have their foundation in commonly received notions than to help the search after truth. So it does more harm than good.

XIII. The syllogism is not applied to the first principles of sciences, and is applied in vain to intermediate axioms; being no match for the subtlety of nature. It commands assent therefore to the proposition, but does not take hold of the thing.

XIV. The syllogism consists of propositions, propositions consist of words, words are symbols of notions. Therefore if the notions themselves (which is the root of the matter) are confused and over-hastily abstracted from the facts, there can be no firmness in the superstructure. Our only hope therefore lies in a true induction. . . .

XIX. There are and can be only two ways of searching into and discovering truth. The one flies from the senses and particulars to the most general axioms, and from these principles, the truth of which it takes for settled and immovable, proceeds to judgment and to the discovery of middle axioms. And this way is now in fashion. The other derives axioms from the senses and particulars, rising by a gradual and unbroken ascent, so that it arrives at the most general axioms last of all. This is the true way, but as yet untried.

---

lor, and he received the further title of Viscount Saint Albans in 1621. Later that year, he was accused of accepting bribes in court cases. He admitted his guilt and apologized profusely, but his public career was over. Banished from court by an act of Parliament, he was imprisoned briefly in the Tower of London. His family's house in London was given to his old ally, the marquis of Buckingham (who would become duke of Buckingham in 1623). Bacon retired to the country house his father had built and married an heiress. He spent his last years making experiments and notes for experiments. He is said to have died of a chill he caught while conducting an experiment with ice.

An expanded Latin version of *Advancement of Learning* appeared in 1623, and the projected volume on natural history appeared posthumously as *Sylva sylvarum* (1627). Bacon never wrote the rest of his masterwork except in fragments, but he left a science-fiction story that suggests what his dream looked like toward the end. In *The New Atlantis* (1627), he imagined a kingdom of science, presided over by a second Solomon. King James was no Solomon; his grandson, Charles II, dabbled in chemistry, however, and became the first patron of the Royal Society. When the society's official history was published in 1667, the frontispiece showed the lord chancellor seated in a room full of books and scientific instruments; at his feet was the motto *artium instaurator*, which may be translated "the restorer of science."

IMPACT

Bacon's *Novum Organum* reversed the accepted Aristotelian methodology of science. Aristotle advocated applying universal rules, known in advance, to specific instances in order to determine their scientific meaning. Bacon, on the other hand, advocated a new empiricism, observing nature in all its manifestations in order to deduce new hitherto unknown rules or principles. The *Novum Organum*, then, is an important part of the age of Scientific Revolution, in which Sir Isaac Newton would deduce the laws of gravitation and the heirs of Nicolaus Copernicus, Johannes Kepler, and Galileo, would establish empirically that the Earth was not the center of the universe. While Bacon's work was not a necessary precursor of any of these other thinkers' triumphs, it was nevertheless a singular and influential expression of a crucial seventeenth century cultural trend, one that informed both the history of science and the broader philosophical Enlightenment of the next century.

See also Heliocentric Universe; Magnetism; Medieval Physics; Scientific Method: Aristotle; Scientific Method: Early Empiricism.

FURTHER READING

Bacon, Francis. *The New Organon and Related Writings*. Edited by Fulton H. Anderson and translated by James Spedding. New York: Liberal Arts Press, 1960.
Eiseley, Loren. *The Man Who Saw Through Time*. Reprint. Boston: Houghton Mifflin, 1973.
Lynch, William T. *Solomon's Child: Method in the Early Royal Society of London*. Stanford, Calif.: Stanford University Press, 2000.
Peltonen, Markku, ed. *The Cambridge Companion to Bacon*. New York: Cambridge University Press, 1996.
Snider, Alvin. *Origin and Authority in Seventeenth-Century England: Bacon, Milton, Butler*. Toronto: University of Toronto Press, 1994.
Webster, Charles. *The Great Instauration*. London: Duckworth, 1975.

Whitney, Charles. *Francis Bacon and Modernity*. New Haven, Conn.: Yale University Press, 1986.

Williams, Charles. *Bacon*. New York: Harper & Bros., 1933.

Wormald, B. H. G. *Francis Bacon: History, Politics, and Science, 1561-1626*. New York: Cambridge University Press, 1993.

—*Thomas Willard*

SCIENTIFIC METHOD: EARLY EMPIRICISM

THE SCIENCE: Early Greek philosophers began to devise theories of the cosmos that abandoned previous mythopoeic explanations, instead depending on observations of nature, or empiricism. This new way of thinking launched a scientific revolution that set the stage for the modern scientific method.

THE SCIENTISTS:

Hesiod (fl. c. 700 B.C.E.), Greek epic poet

Thales of Miletus (c. 624-c. 548 B.C.E.), Greek philosopher and scientist

Anaximander (c. 610-c. 547 B.C.E.), Greek philosopher often called the founder of astronomy

Anaximenes of Miletus (early sixth century-latter sixth century B.C.E.), Greek philosopher and scientist

Xenophanes (c. 570-c. 478 B.C.E.), Greek philosopher and poet

Pythagoras (c. 580-c. 500 B.C.E.), Greek philosopher, astronomer, and mathematician

Heraclitus of Ephesus (c. 540-c. 480 B.C.E.), Greek philosopher known for his book *On Nature*

Parmenides (c. 515-after 436 B.C.E.), Greek philosopher of metaphysics

Democritus (c. 460-c. 370 B.C.E.), Greek philosopher associated with atomism

MYTH AND SCIENCE

Before the sixth century B.C.E., human beings everywhere explained the world in mythological terms. These myths depicted humankind dependent on the wills of inscrutable forces, envisioned as gods, that created the world and acted on whim. The prelogical mentality of early people understood the forces of nature to possess powerful consciousnesses similar to human will. No other explanation or scientific foundation on which to build a different understanding of the world and nature yet existed.

Similarly, most Greeks honored the epic poets Homer (early ninth century-late ninth century B.C.E.) and Hesiod as their teachers. Hesiod's *Theogony* (c. 700 B.C.E.; English translation, 1728) is the earliest Greek version of the origins of the cosmos. The Greek term *kosmos* (cosmos) refers to the organized world order. In Hesiod's account, the origin of all things was *chaos*, formless space or a yawning watery deep, the opposite of cosmos. In time there emerged, either independently or by sexual union, Gaia (Earth), Tartaros (Hades), Eros (Love), Night, Day, and Aither (upper air), Sea, and Ouranos (Sky), and boundless Okeanos (Ocean). A generation of powerful Titans was engendered, and finally the Olympian gods descended from Ouranos and Gaia.

A New Worldview

About 600 B.C.E., in Ionia (western Turkey), a new way of perceiving the world was beginning. Confronted by the confusing mythologies of ancient Near Eastern peoples, their own no better, a handful of Greeks over three generations attempted to explain the origins and components of the seen world without mythology. Their great discovery was that to one seeking knowledge—the philosopher—the world manifests internal order and discernible regularity. Nature can be understood. The world is a cosmos.

From allusions in Homer and Hesiod came hints. The sky was thought to be a metallic hemispheric bowl covering the disk of Earth. The lower space immediately above the disk was *aër*, breathable air; the upper part of the bowl-space was *ouranos* or *aither*. Below its surface, the Earth's deep roots reached down to Tartaros, the deepest part of Hades (the underworld realm of the dead), as far below Earth as sky is above it. *Okeanos* (ocean), infinitely wide, encircled the disk of Earth and was the source of all fresh and salt waters. Such a mixture of the empirical and the imaginative was common to most mythopoeic cosmologies.

Thales of Miletus

Thales of Miletus was the first to rationalize the myths. He conceived the Earth-disk as floating on the ocean and held the single substance of the world to be water. His reasoning, according to Aristotle, was that water can be gaseous, liquid, and solid; life requires water; Homer had surrounded the Earth by Okeanos. As a unified source of all things, Thales' choice of water was a good guess, but it begged for alternatives. More important, in reducing multiple things to water, Thales had taken a first step in establishing inductive reasoning (from particular examples to general principles) as a scientific methodology.

ANAXIMANDER

Anaximander, companion of Thales, was a polymath: astronomer, geographer, evolutionist, philosopher-cosmologist. It is nearly impossible to do justice to his intellectual achievement. He was the first Greek to write in prose. He said that animal life began in the sea and that humans evolved from other animals. He made the first world map, a circle showing Europe and Asia plus Africa equal in size, all surrounded by ocean. Anaximander's cosmos was a sphere with a drum-shaped earth floating in space at its center. The Sun, stars, and Moon revolved around the Earth, seen through openings in the metallic dome of the sky.

In place of Thales' water, Anaximander offered *apeiron*, an eternal, undefined, and inexhaustible basic stuff from which everything came to be and to which everything returns—a sophisticated chaos. Convinced by his own logic, Anaximander imputed an ethical necessity to this process. Things coming to be and claiming their share of *apeiron* thus deprive others of existence. In his words, "they must render atonement each to the other according to the ordinances of Time." This eternal process operates throughout the cosmos. Using terms such as *kosmos* (order), *diké* (justice), and *tisis* (retribution), Anaximander enunciated the exalted idea that nature itself is subject to universal moral laws.

ANAXIMENES OF MILETUS

The contributions of Anaximenes of Miletus pale before those of Anaximander. What best defines Anaximenes is his empirical approach. He posited air as the primal stuff that gives rise to all things. Observing air condensing into water, he conceived a maximum condensation of air into stone. Similarly, by rarefaction, air becomes fire or soul. The Earth and other heavenly bodies, being flat, ride on air in its constant motion.

XENOPHANES

Xenophanes, an Ionian who had moved to Italy, represents a new generation of thinkers. He was a skeptic who trusted only his own observations about the world. He interpreted the new natural explanations of the universe that had challenged the older Hesiodic mythopoeic construct as the abandonment of the old, often immoral, anthropomorphic gods, who dressed in clothes and spoke Greek. He posited a single spiritual creator-god who controls the universe without effort, by pure thought. In this monotheism, Xenophanes was alone among the Greeks.

Insightfully, Xenophanes said human knowledge about the universe is limited and the whole truth may never be known. He taught that natural events have natural, not divine, causes. The rainbow is only a colored cloud. The sea is the source of all waters, winds, and clouds. From sea fossils found in rocks, his cosmogony deduced a time when land was under water. Civilization was the work of men, not gods.

PYTHAGORAS

Pythagoras, an Ionian mathematician in southern Italy, had noticed that the sounds of lyre strings varied according to their length and that harmonies were mathematically related. He saw that proportion can be visually perceived in geometrical figures. From these notions he and his followers described a cosmos structured on a mathematical model. Instead of adopting Anaximander's "justice" or Heraclitus of Ephesus's *logos* (a sort of primordial reason or order identified with speech or the word) as the dominant organizing principle, the Pythagoreans preferred numerical harmony. Pythagoras thus added a dimension to the ancient concepts of due proportion and the golden mean that pervaded Greek thought. These concepts are seen in Greek sculpture and architecture and as moral principles in lyric and dramatic poetry and historical interpretations, where *hybris* (hubris, or excess) and *sophrosyné* (moderation) were fundamental principles of human behavior.

Inevitably, Greek physical philosophy began to investigate the process of knowing. Number is unchanging; ten is always ten. In a world of apparently infinite diversity and flux, numbers, as opposed to objects of experience, can

Pythagoras. (Library of Congress)

be known perfectly. Although the Pythagoreans went too far in trying to explain everything by numbers, they taught that a nature based on mathematical harmony and proportion was knowable.

HERACLITUS

Heraclitus argued that change, though sometimes imperceptible, is the common element in all things. All change, he said, takes place along continuums of opposite qualities, such as the hot-cold line or dry-moist line. His contribution, however, was his idea of *logos* as the hidden organizing principle of the cosmos. *Logos* maintains a protective balance (the golden mean again) among all the oppositional tensions in the world.

PARMENIDES AND DEMOCRITUS

Parmenides and Democritus contributed logic to the Greek discovery of the cosmos. In the mid-fifth century, Democritus reasoned to a world built of the smallest thinkable indivisible particles, atoms. Parmenides—struck by the constant flux of the physical world and seeking, as Pythagoras, an unchanging object of knowledge that mind can grasp—saw existence, or Being, as the common element of things in the cosmos. He proposed the logic that while things change, Being itself cannot change, for nothing and no place exists outside of the sphere of Being, so nothing could enter or leave. He is thus the most metaphysical of the philosophers, initiating ideas that would only be completed by Plato and Aristotle, the greatest of the philosophers.

IMPACT

The significance of the Ionian philosophers is that, within little more than a century after breaking with mythopoeic interpretations of the world, they had asserted its atomic makeup, conceived human evolution, discovered induction and logic, and practiced a curiosity about all natural phenomena. This was one of history's great intellectual revolutions—the origins of scientific speculation.

See also Galen's Medicine; Greek Astronomy; Greek Medicine; Scientific Method: Aristotle; Scientific Method: Bacon.

FURTHER READING

Kirk, G. S., and J. E. Raven. *The Presocratic Philosophers*. 2d ed. New York: Cambridge University Press, 1984.

Long, A. A. *The Cambridge Companion to Early Greek Philosophy*. New York: Cambridge University Press, 1999.

Morgan, Kathryn A. *Myth and Philosophy from the Presocratics to Plato*. New York: Cambridge University Press, 2000.

Popper, Karl R. *The World of Parmenides: Essays on the Presocratic Enlightenment*. Edited by Arne F. Petersen and Jørgen Mejer. New York: Routledge, 1998.

Ring, Merrill. *Beginning with the Presocratics*. 2d ed. New York: McGraw-Hill, 1999.

Wakefield, Robin, ed. *The First Philosophers: The Presocratics and the Sophists*. New York: Oxford University Press, 2000.

—*Daniel C. Scavone*

SEAFLOOR SPREADING

THE SCIENCE: Harry Hammond Hess's idea of seafloor spreading as the reason for continental drift had the same impact on geology that Charles Darwin's evolution theory had on biology.

THE SCIENTISTS:
Harry Hammond Hess (1906-1969), American geologist
Alfred Lothar Wegener (1880-1930), German scientist-explorer
Robert S. Dietz (1914-1995), American geologist
Matthew F. Maury (1822-1891), U.S. Navy officer and oceanographer

A GEOPHYSICAL "FAIRY TALE"

The Princeton University professor Harry Hammond Hess is noted for his scientific contributions to the field of geology, specifically for his groundbreaking *History of the Ocean Basins* (1962), in which he proposed seafloor spreading as the long-sought-after mechanism for Alfred Lothar Wegener's theory of continental drift, which he had proposed fifty years before. The elements of seafloor spreading, the splitting of the original Pangaea supercontinent into several continental-size plates, and movement of those plates to their present positions are collectively known as "plate tectonics."

Yet the hostility of the geologic community toward previous seafloor spreading hypotheses kept Hess from publishing his theory. Robert S. Dietz, working for the Navy, published virtually identical ideas and coined the phrase "seafloor spreading" in a 1961 article, "Continent and Ocean Basin Evolution by Spreading of the Seafloor," published in *Nature*.

World War II (1939-1945) delayed research into the question, and from the 1930's to the mid-1950's, continental drift remained a theory held with great passion by a minority of geologists. As late as 1966, the University of Hamburg physicist Pascual Jordan described the theory as the geophysicists' "favorite fairy tale."

THE MID-ATLANTIC RIDGE

Postwar advances in technology, in methodology, and in the new science of paleomagnetism (the study of the direction and intensity of the Earth's magnetic field through geologic time) lent support to Wegener's theory. The U.S. Navy directed its interest to the ocean floor, and other seagoing nations also initiated active research programs that led to the International Geophysical Year (July, 1957, to December, 1958), the first multinational research effort. This effort focused on almost every area of geologic research and led scientists to realize that the Earth, particularly the ocean, was very different from what they had previously imagined. One of the most curious features was the Mid-Atlantic Ridge, which spans the Atlantic Ocean from north to south. Understanding the feature led to an understanding of plate tectonics.

Matthew F. Maury, director of the U.S. Navy's Department of Charts and Instruments, had first recognized the Mid-Atlantic Ridge in 1850 while measuring ocean depths aboard the USS *Dolphin*. Maury named it the "Dolphin Rise" and published a map of it in his *The Physical Geography of the Sea* (1855). Data from the HMS *Challenger* expedition (1872-1876) supplemented Maury's map, but the details of the Mid-Atlantic Ridge remained vague. In 1933, German oceanographers Theodor Stocks and Georg Wust produced the first detailed map of the ridge, noting a valley that seemed to be bisecting it. Later, in 1935, geophysicist Nicholas H. Heck found a strong correlation between earthquakes and the Mid-Atlantic Ridge.

The idea of a seismically active ridge received further support in 1954. That year Jean P. Rothé, director of the International Bureau of Seismology in Strasbourg, mapped a continuous belt of earthquake epicenters from Iceland through the mid-Atlantic around South Africa, through the Indian Ocean, and on to the African Rifts and the Red Sea. In 1956, Maurice Ewing

HESS'S THEORY OF CONVECTION

Trench

Mid-Ocean Ridge
(Divergent Plate Boundary)

Trench

← Plate   Plate →

Upwelling

Convection Cell   Convection Cell

Mantle

In 1960, Harry Hammond Hess proposed that seafloor spreading, powered by convection currents within the Earth's mantle, might be the cause of continental "drift." The idea of continental drift became the theory of plate tectonics, which is accepted by most earth scientists today.

and Bruce C. Heezen continued the German technique of echo sounding at the Lamont Geological Observatory and found that the Mid-Atlantic Ridge was more than 64,000 kilometers long and, more important, it had a rift valley along the entire crest.

In 1961, Ewing and Mark Landisman discovered that this ridge system extends throughout all the world's oceans, is seismically and volcanically active, and is mostly devoid of sediment cover. The question of whether the ridge system was covered with sediment—and the amount and age of that sediment—was important: It would reveal clues to the age of the ridge itself. Ivan Tolstoy and Ewing first characterized the sediment cover in 1949, describing a main ridge of thin sediment and flanks of thick sediment. The age of the sediments increased as one moved from the ridge toward the continents, the oldest age being only about sixty-five million years old.

Central to the interpretation of this underwater mountain range was the early 1950's paleomagnetic research of Patrick M. S. Blackett and his student, Keith Runcorn, at the University of Manchester. Their studies of fossil magnetism suggested that in the geologic past, the inclination, the declination, and even the polarity of the Earth's magnetic field had been very different from current orientations. The seemingly chaotic data formed a consistent pattern only upon assuming that the continents had moved relative to the magnetic poles and to one another. Magnetic studies of the seafloor by other oceanographers revealed a symmetrical, zebralike pattern about the midoceanic ridge in 1957.

A Bold New Synthesis

The period between 1960 and 1965 was one of great uncertainty and multiple directions for geologists. In 1960, Hess synthesized the oceanic data of the 1950's into a bold new theory. Hess's theory was so novel and radical that he did not attempt to publish it in the usual professional journals but included it in a 1960 report to the Office of Naval Research. Hess also widely circulated reprints among his colleagues.

In his 1960 report, Hess proposed that the midoceanic ridges were the locations of upwelling "mantle convection cells": that is, areas of the Earth's mantle that progressively moved the seafloor outward from the ridge and eventually under the continents. The mantle is part of the molten core of the Earth. Hess suggested that different parts of it (cells) may spin like wheels, driven by changes in temperature (convection) between the lower and upper parts of the mantle.

Impact

This driving mechanism brought together the divergent data of post-World War II research into one coherent theory. It explained the rift valley in the middle of the ridge, the correlation of the ridge with earthquake epicenters, the continuation of the ridge throughout the oceans, the thin sediment in the middle of the ridge and its thickening toward the edges, and the symmetrical paleomagnetic zebra patterns. In addition, the energy of the mantle convection currents was sufficient to drive the continents.

In 1966, in recognition of his scientific breakthrough, Hess received the Geological Society of America's Penrose Medal, the geologist's equivalent of the Nobel Prize. By 1967, seafloor spreading was the dominant theory, and virtually all earth scientists began to reinterpret their data in the light of the new theory.

See also Continental Drift; Earth's Core; Earth's Structure; Geomagnetic Reversals; Hydrothermal Vents; Mid-Atlantic Ridge; Plate Tectonics; Radiometric Dating.

Further Reading

Engel, A. E. J., H. L. James, and B. F. Leonard, eds. *Petrologic Studies: A Volume in Honor of A. F. Buddington.* New York: Geological Society of America, 1962.

James, H. L. "Harry Hammond Hess." *National Academy of Sciences Biographical Memoirs* 43 (1973): 108-128.

Le Grand, Homer E. *Drifting Continents and Shifting Theories.* New York: Cambridge University Press, 1988.

Scientific American. *Continents Adrift.* San Francisco: W. H. Freeman, 1973.

Sullivan, Walter. *Continents in Motion: The New Earth Debate.* New York: McGraw-Hill, 1974.

Young, Patrick. *Drifting Continents, Shifting Seas.* New York: Impact Books, 1976.

—*Richard C. Jones and Anthony N. Stranges*

SMALLPOX VACCINATION

THE SCIENCE: English physician Edward Jenner was the first person to establish the scientific legitimacy of smallpox vaccinations through his experiments and research publications. His campaign to popularize the procedure led to its worldwide use and effectively protected millions from an often fatal disease.

THE SCIENTISTS:

Edward Jenner (1749-1823), English physician

James Phipps (1788-1808), first person to be vaccinated by Edward Jenner

Benjamin Jesty (1736-1816), English farmer who vaccinated his family against smallpox in 1774

William Woodville (1752-1805), head of the London Smallpox and Inoculation Hospital

George Pearson (1751-1828), physician at St. George's Hospital in London

VARIOLATION

In eighteenth century England, smallpox was a leading cause of death, and traditional methods of treating it were largely ineffective. The practice of variolation was introduced to England from the Ottoman Empire in 1721 and gained general acceptance after some successful trials. This procedure involved inoculating patients with puss from smallpox sores in hopes of giving them a mild case of the disease and future immunity. However, the risks of a patient developing a serious, possibly lethal, case of smallpox and even creating an epidemic were significant, and there was a clear need for a safer and more effective method of protection from the disease.

Jenner and Cowpox

Edward Jenner, a physician in Berkeley, England, in the county of Gloucestershire, began variolating patients using a refined procedure developed by Robert Sutton in 1768. Jenner found that his patients who in the past had contracted cowpox, a relatively mild disease, did not react to the smallpox virus. This finding was consistent with the conventional wisdom in rural areas that cowpox conferred an immunity to smallpox, which had been supported in reports to the Medical Society of London in the mid-1760's by several physicians, including at

Edward Jenner. (Library of Congress)

least two from Gloucestershire. In fact, a farmer in Yetminster, England, named Benjamin Jesty successfully protected his wife and two sons from a smallpox epidemic by vaccinating them with the puss from the udders of cows suffering from cowpox in 1774. Jenner, however, always maintained that he was unaware of these earliest documented smallpox vaccinations.

By the early 1780's, Jenner's interest in the connection between cowpox and smallpox immunity led him to distinguish between two similar but distinct diseases, "spontaneous," or genuine, cowpox, which created an immunity from smallpox, and "spurious," or false, cowpox, which did not. In May of 1796, a young woman named Sarah Nelmes came to Jenner to be treated for cowpox. On May 14, Jenner vaccinated James Phipps, an eight-year-old boy, by placing fluid from a sore on Nelmes' hand into two small incisions on the boy's arm. A week later, Phipps developed the symptoms of cowpox, including infected sores, chills, head and body aches, and loss of appetite. The child recovered quickly, and, on July 1, 1796, Jenner variolated Phipps using fluid from smallpox pustules, and he had no reaction. Jenner inoculated the boy several more times in this manner with the same results.

Arm to Arm

In late 1796, Jenner submitted a paper to be considered for publication in *Philosophical Transactions of the Royal Society*, England's premiere scientific journal. The manuscript described the cases of thirteen former cowpox sufferers who exhibited no reaction when variolated by Jenner, as well as his experiments with Phipps. The Council of the Royal Society rejected the article, and berated Jenner in scathing terms, characterizing his findings as being unbelievable, and "in variance with established knowledge," and advising him that advancing such wild notions would destroy his professional reputation. Jenner was undaunted and began experimenting again in the spring of 1798, when cowpox broke out again in Gloucestershire. Through these studies he learned that cowpox could be transferred from one patient to another by using the puss from the sores of one vaccinated person to vaccinate another, and so forth. This discovery of "arm-to-arm vaccination" made a natural outbreak of cowpox unnecessary as a source of vaccine.

In June of 1798, Jenner independently published the findings from all of his research to date, including reports of the cases from his first manuscript and nine other patients he had vaccinated beside Phipps. This seventy-five-page book was titled *An Inquiry into the Causes and Effects of the Variolae Vaccinae, a Disease Discovered in Some of the Western Counties of England, Particularly Gloucestershire, and Known by the Name of the Cow Pox*. The word *variolae* is "smallpox" in Latin, and *vaccinae* is from *vaca*, which is Latin for "cow." In his inquiry, Jenner described the process now called "anaphylaxis," the body's allergic reaction to a foreign protein after a previous exposure, and coined the term "virus" to describe the mechanism of cowpox transmission.

THE FIGHT FOR VACCINATION

The London medical establishment's initial reaction to Jenner's publication was extremely negative. Just as in 1796, some prominent physicians questioned the validity of Jenner's findings. Others, who were profiting handsomely from variolation, attacked Jenner for fear of losing their lucrative monopoly on protecting the public from smallpox. Jenner had rejected the suggestion that he could become personally wealthy from his discovery, and he planned to share it with all of England and the world. After the publication of his findings Jenner tried for three months to find people who would agree to be vaccinated in order to demonstrate the effectiveness of the procedure. He did not find a single volunteer because of the public attacks on his professional competence.

Instead, Jenner pursued his goal of popularizing vaccination indirectly, through London physicians to whom he provided vaccine. For example, the director of the London Smallpox and Inoculation Hospital, William

Woodville, vaccinated some six hundred people in the first half of 1799. Based on vaccinations that he performed in 1799, George Pearson of St. George's Hospital replicated Jenner's findings, and tried to take credit for the procedure. Woodville, who caused several cases of smallpox and at least one death by inadvertently contaminating some vaccine with the smallpox virus, blamed Jenner's procedure in order to protect his own reputation. However, a nationwide survey conducted by Jenner that documented cases of immunity to smallpox by former cowpox sufferers, clearly validated his work.

By late 1799, vaccination had gained widespread acceptance, and the procedure was being performed not only by physicians, but also by schoolteachers, ministers, gentlemen farmers, and others in all parts of the country. Jenner continued to report the results of his research on vaccination through publications such as *The Origin of the Vaccine Inoculation* (1801). In recognition of his achievements, Parliament awarded Jenner £10,000 in 1802 (the equivalent of more than $500,000 today) and an additional £20,000 in 1807. Oxford, Harvard, and Cambridge Universities honored him as well.

IMPACT

Edward Jenner's work on refining and promoting the use of smallpox vaccinations, before the development of antibiotics, was a major breakthrough in preventive medicine. Countless lives were undoubtedly saved in Great Britain during the years immediately following Jenner's efforts, given the high mortality rates during earlier smallpox epidemics. Furthermore, his successful lobbying for a government-sponsored national vaccination program eventually led to the passage of the Vaccination Act in 1840, which provided for the free vaccination of infants and made variolation illegal. Subsequent laws made vaccination mandatory, with severe penalties for noncompliance. By 1871, 97.5 percent of England's population reportedly had been vaccinated.

Jenner's method of preserving vaccine for up to three months enabled him to share his vaccination procedure with the world. As a result, an estimated 100,000 people had been vaccinated worldwide by the end of the eighteenth century. Shortly thereafter, Benjamin Waterhouse, a professor at the Harvard School of Medicine, used vaccine from England to perform the first vaccinations in the United States on his young son and servants. Jenner also shipped vaccine to President Thomas Jefferson, who had eighteen of his relatives vaccinated and established the National Vaccine Institute, with Waterhouse as its director, to spread vaccination throughout the country. In addition, mass vaccination programs were initiated in all Spanish colonies in North and South America and Asia by King Charles IV of Spain, in India by the British governor general, for the French army by Napoleon, and in numerous other countries. These programs were all undertaken in the early 1800's using Jenner's vaccine.

By 1967, although smallpox had completely disappeared from North America and Europe, there were still 10-15 million cases reported in the world annually. The World Health Organization initiated an effort to eradicate smallpox worldwide. The campaign was declared a success in 1980. Jenner's work is credited not only with the defeat of smallpox but also with helping to establish the science of immunology, which has produced vaccines against numerous lethal and debilitating diseases.

See also Anesthesia; Antisepsis; Aspirin; Contagion; Diphtheria Vaccine; Germ Theory; Hybridomas; Immunology; Penicillin; Polio Vaccine: Sabin; Polio Vaccine: Salk; Schick Test; Streptomycin; Vitamin C; Vitamin D; Yellow Fever Vaccine.

FURTHER READING

Barquet, Nicolau, and Pere Domingo. "Smallpox: The Triumph over the Most Terrible of the Ministers of Death." *Annals of Internal Medicine* 127 (1997): 635-642.
Baxby, Derrick. "The End of Smallpox." *History Today*, March, 1999: 14-16.
Bazin, Hervé. *The Eradication of Smallpox: Edward Jenner and the First and Only Eradication of a Human Infectious Disease.* Translated by Andrew Morgan and Glenise Morgan. San Diego, Calif.: Academic Press, 2000.
Fisher, Richard B. *Edward Jenner, 1749-1823.* London: Andre Deutsch, 1991.
Kerns, Thomas A. *Jenner on Trial: An Ethical Examination of Vaccine Research in the Age of Smallpox and the Age of AIDS.* Lanham, Md.: University Press of America, 1997.
Plotkin, Susan L., and Stanley A. Plotkin. "A Short History of Vaccination." In *Vaccines*, edited by Stanley A. Plotkin and Walter A. Orenstein. Philadelphia: W. B. Saunders, 2004.

—*Jack Carter*

SOLAR WIND

THE SCIENCE: Eugene N. Parker predicted the existence of the solar wind, which was confirmed by a Soviet satellite in 1959.

THE SCIENTISTS:

Eugene N. Parker (b. 1927), American physicist
Ludwig Biermann (1907-1987), German astrophysicist
Sydney Chapman (1889-1970), English mathematician and physicist
Subrahmanyan Chandrasekhar (1910-1995), Nobel laureate in physics
K. I. Gringauz (b. 1925), the principal investigator for the Soviet
    satellite that detected the existence of the solar wind

A STREAM OF CHARGED PARTICLES

When the idea of the solar wind was first proposed, the notion of a
steady stream of charged particles emanating from the Sun at supersonic
speeds was hard to accept. Even when the solar wind was confirmed by satellite, it continued to be dismissed by some as impossible. Some events that
are now known to be caused by the solar wind were quite familiar by the
1950's. For example, the auroras, both at the North Pole (aurora borealis) and
at the South Pole (aurora australis), had been observed for centuries. The fact
that the tail of a comet always points away from the Sun, no matter in which
direction the comet is moving, was known. Magnetic storms, which affect
the Earth's magnetic field and induce voltages in telegraph and power
lines, had been observed. For each of these occurrences, scientists knew
that the Sun was responsible, or at least involved, but did not know how.

In 1957, Eugene N. Parker was an assistant physics professor at the University of Chicago. He had been studying the origin of Earth's magnetic
field, the atmosphere of the Sun, and cosmic rays. Ludwig Biermann, director of the Max Planck Institute for Astrophysics in Munich, was visiting
the University of Chicago that same year. Biermann told Parker of his studies of comet tails. Comets are essentially chunks of rock dust and ice, "dirty
snowballs." As a comet nears the Sun, solar heat vaporizes the ice, which
releases dust particles, and both vapor and dust create a tail. Solar radiation ionizes the atoms in the tail, making it one of the most spectacular
sights in the evening sky.

TAILING THE COMET

Biermann was interested in why comet tails always point away from
the Sun, even when that means that the tail is pointing in the same direction in which the comet is moving. It was thought that the Sun's electromagnetic field exerted radiation pressure on the comet, thereby pushing
the tail away from the Sun. Astronomers also believed that this pressure,
while very small, was still stronger than the tail of a comet.

Yet the dust and gas in the tail were not merely being carried away from the comet—they were being blown away with great force. Further research by Biermann showed that a comet's tail did not have enough surface area for solar radiation to have that effect. He concluded that there was only one other explanation: "solar corpuscular radiation," the discharge of particles from the Sun at the time of a "solar flare" (the sudden eruption of hydrogen gas on the Sun's surface). This corpuscular radiation evidently was shot out from the Sun at an average velocity of 500 kilometers per second. Such bursts were known to cause auroras and magnetic storms. The rest of the time, however, interplanetary space was thought to be empty.

Parker was influenced by Biermann's theories. Taking hold of the observation that the tail of a comet always points away from the Sun, the fact that auroras are always present, and other cosmic observations, Parker agreed that interplanetary space must continually be filled with solar corpuscular radiation. Now he needed to determine why it was there and why it was moving so forcefully.

Shortly after a discussion with Biermann, Parker was in Boulder, Colorado, where he had been invited to give a lecture at the High Altitude Observatory. He had an opportunity to learn of the work of Sydney Chapman on the solar corona. The corona is an envelope of thin, hot gas that surrounds the Sun. Chapman showed Parker some calculations indicating that the outer atmosphere of the Sun, the corona, extends out into space, past the orbit of Earth. The high temperature of the corona was causing it to

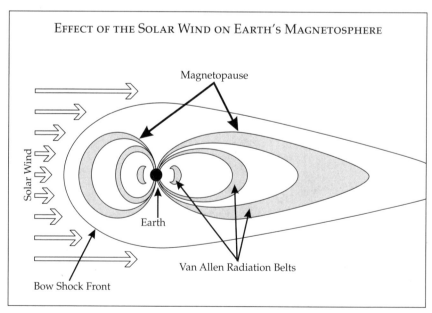

EFFECT OF THE SOLAR WIND ON EARTH'S MAGNETOSPHERE

Magnetopause

Solar Wind

Earth

Bow Shock Front

Van Allen Radiation Belts

expand slowly upward against the gravitational field of the Sun. It gradually increased in speed until it reached supersonic velocity.

Chapman's conclusion not only was novel and interesting but also appeared to be inescapable. A little more thought, however, seemed to indicate a conflict with Biermann's equally inescapable conclusion that solar corpuscular radiation continually fills interplanetary space. The two could not possibly exist together. Parker, however, believed that solar corpuscular radiation and the extended solar corona described the same phenomenon. They both were, as he came to name it, the "solar wind."

In 1958, Parker wrote a paper titled "Dynamics of the Interplanetary Gas and Magnetic Fields." He reconciled the work of Chapman and Biermann and included equations showing supersonic velocities for the solar wind of several hundred kilometers per second. With the publication of Parker's paper in the *Astrophysical Journal*, all that remained to be accomplished was an actual "sighting." Instruments designed by Soviet scientist K. I. Gringauz and carried aboard a Luna satellite were the first to detect a gas moving past the Earth faster than 60 kilometers per second—a supersonic velocity that confirmed Parker's prediction.

IMPACT

The discovery of the solar wind has given the correct explanation for the auroras, magnetic disturbances, and the behavior of comet tails and has added to our knowledge of stars. Most stars have their own stellar winds, similar to the solar wind.

Traditionally, stars have been seen as tranquil objects, shining for billions of years and eventually dying. It has been learned that a star is very active. For example, it has been established that the luminosity of the Sun varies by one part in six hundred, and there is evidence from other Sun-type stars that it could vary at times by one part in one hundred, in the space of a few years. A decrease in luminosity of that magnitude would cause the polar ice caps to advance, thus producing a small "ice age." In fact, there is evidence that the Sun's luminosity has fluctuated recently: The Little Ice Age began in the thirteenth century and ended in the middle of the eighteenth century. The Little Ice Age affected global agricultural output, leading to hardship in China and in Europe. Killing frosts in the North American Great Plains were commonplace each summer.

Understanding the solar wind shows scientists that the Sun is losing mass at the rate of about 1 million tons per second. That, however, is not a problem. Because of the immense size of the Sun, the loss it has experienced has amounted to only one ten-thousandth of its original mass.

See also Cassini-Huygens Mission; Extrasolar Planets; Halley's Comet; Magnetism; Nebular Hypothesis; Oort Cloud; Planetary Formation; Pluto; Stellar Evolution; String Theory; Van Allen Radiation Belts; Very Long Baseline Interferometry; X-Ray Astronomy.

FURTHER READING

Menzel, Donald H., and Jay M. Pasachoff. *Stars and Planets*. Boston: Houghton Mifflin, 1983.
Mitton, Simon, ed. *The Cambridge Encyclopaedia of Astronomy*. New York: Crown, 1977.
Muirden, James. *The Amateur Astronomer's Handbook*. 3d ed. New York: Harper & Row, 1987.
The Rand McNally New Concise Atlas of the Universe. New York: Rand McNally, 1989.
Zeilik, Michael. *Astronomy: The Evolving Universe*. 2d ed. New York: Harper & Row, 1979.

—*John M. Shaw*

SPACE SHUTTLE

THE SCIENCE: Circling above the globe, the world's first reusable spacecraft opened the future to payloads and experimenters to whom space was previously inaccessible.

THE ASTRONAUTS:
Joe H. Engle (b. 1932), U.S. Air Force colonel and STS 2 commander
Richard H. Truly (b. 1937), Navy captain and STS 2 pilot
John W. Young (b. 1930), STS 1 commander who became the first person to fly into space six times when he became commander of the STS 9 mission
Robert L. Crippen (b. 1937), Navy captain and STS 1 pilot

FUNDING AND DEVELOPMENT

After the success of the lunar landings in 1969, the National Aeronautics and Space Administration (NASA) wanted to send astronauts to Mars, build a fifty-person Earth-orbiting space station serviced by a reusable ferry (space shuttle), and build a second space station orbiting the Moon.

THE SHUTTLE CONFIGURED
FOR LAUNCH

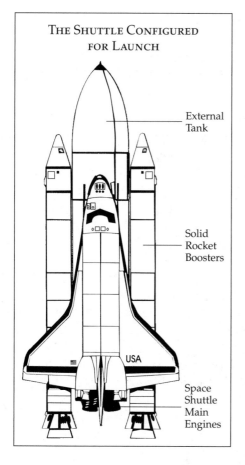

External
Tank

Solid
Rocket
Boosters

USA

Space
Shuttle
Main
Engines

At $8 to $10 billion per year, Congress rejected the proposal as too expensive. NASA proposed two other programs, each one simpler and less expensive, but Congress refused to finance either of them. By the spring of 1971, NASA was determined to secure appropriations for at least the space shuttle, which would be the first step if any of the other programs were approved.

NASA turned to the Department of Defense (DOD). Under NASA's plan, the DOD would provide a large portion of the funding for the shuttle; in return, the shuttle would be used for military as well as scientific missions. The shuttle also would become the only launch vehicle in NASA's fleet, thereby replacing expendable boosters and saving billions of dollars. Congress approved this proposal, and plans were drawn for a shuttle system.

Budgetary cuts forced NASA to reduce the shuttle program—formally known as the Space Transportation System (STS)—to the four-part system it became. Basically, the shuttle system consists of an orbiter, to which are attached three main engines to be used during launch; a large, external fuel tank for the engines; and a pair of strap-on solid rocket boosters to get the stack off the launch pad. The orbiter could be reused, as could the major components of the solid rocket boosters, which would parachute back into the ocean after running out of fuel. Only the external fuel tank would have to be discarded.

By 1977, a series of drop tests were done using an orbiter that was identical to later flight versions but that was incapable of spaceflight. *Enterprise* (named for the starship from the television series *Star Trek*) was placed on top of a modified Boeing 747 airliner and released. This allowed the glide characteristics of the orbiter to be determined and gave shuttle pilots hands-on experience with landing at speeds close to those expected on orbital missions.

Can They Do It Twice?

On December 29, 1980, the first operational space shuttle, *Columbia*, was rolled to Launch Complex 39A at the Kennedy Space Center in preparation for its first flight. This was to be a piloted flight, and it marked the first time that a piloted space vehicle would be flown without the benefit of robotic test flights in space—another result of budgetary limits.

The STS 1 mission blasted off on April 12, 1981, twenty years to the day after the first astronaut, Yuri Gagarin, had been launched into space by the Soviet Union. The STS 1 commander was John W. Young; the pilot was Robert L. Crippen. The two-day "shakedown cruise" showed that the systems worked and that the orbiter could convert from rocket to orbiter to glider without enormous problems. The only major concept yet to be tested was that of reusability. The only way to find out was to fly *Columbia* a second time.

The STS 2 mission was scheduled to be launched on November 4, 1981. The STS 2 commander was Joe H. Engle; the pilot was Richard H. Truly. A problem with one of *Columbia*'s auxiliary power units (APUs), however, could not be corrected in time to meet the launch deadline for the day. The launch was pushed back to November 12. At 59 seconds past 10:09 a.m. eastern standard time, the era of the reusable shuttle vehicle began. The launch proceeded normally, and *Columbia* was placed into a 222-kilometer-high orbit above Earth. The mission was supposed to last five days but was shortened to a little more than two days after a fuel cell failure less than five hours into the mission. During the shortened mission, more than 90 percent of the high-priority flight tests were completed successfully. The Development Flight Instrumentation, used to monitor *Columbia*'s systems during the flight, showed that the orbiter functioned as planned. The Remote Manipulator System's 15-meter robot arm was first flown on this mission. On later missions, it would be used to handle large payloads.

Columbia landed at Edwards Air Force Base on November 14 at 6:23 p.m. Pacific time. The space shuttle had been proved to be reusable.

Impact

In a throwaway society where nearly everything is disposable, the idea of building a spacecraft that could be used many times over was pure science fiction until the space shuttle. Prior to the Space Transportation System, a satellite or probe was launched into space and, if it arrived at its destination successfully, kept operating until it ran out of fuel or lost electrical power. Then it was discarded for a newer model.

Space shuttle Atlantis *takes a ride home after landing at Edwards Air Force Base in 1998.* (NASA/DFRC)

By building a vehicle that could carry large payloads to and from low-Earth orbit, it was possible to retrieve those old satellites and either repair them or bring them back to Earth. Doing so would save the space program a great deal of money. The ideal and most economical craft for this job was one that reused all or most of its parts: the space shuttle.

The space shuttle program continued to enjoy a number of achievements in the years that followed *Columbia*'s historic flight. Sadly, the entire U.S. space program was dealt a severe blow with the fiery explosion of the *Challenger* spacecraft and the tragic loss of its entire crew, including the first civilian astronaut, on January 28, 1986. NASA identified the problem and corrrected it, returning to flight in 1988 with STS-26 and completing many more missions—increasingly to service the growing International Space Station (ISS). When a second accident killed the seven-member crew of *Columbia* on February 1, 2003, during the reentry of STS-107, NASA was forced into another hiatus to determine what had caused small pieces of the heat-resistant foam tiles to peel from the shuttle upon liftoff and again put measures in place to correct the problem: The return-to-flight mission, STS 114, lifted off in July of 2005 with unprecedented cameras and procedures in place to inspect the orbiter in space as it also delivered supplies to the ISS.

As the shuttle fleet aged and with two of the orbiters lost, NASA continued its plans to complete the shuttle's obligations to the ISS. It also made plans for a new fleet of crew transfer vehicles to replace the shuttle program, whose lifetime was expected to end around 2010. In the meantime— more than a quarter century and many successful launches later—the space shuttle remained NASA's primary source for piloted space exploration. Counted among its many successes are the deployment of the Hubble Space Telescope and Chandra X-Ray Observatory; the Galileo, Ulysses, and Magellan probes; and the early stages of construction of the International Space Station.

See also Cassini-Huygens Mission; Earth Orbit; Galileo Mission; International Space Station; Mars Exploration Rovers; Moon Landing; Voyager Missions; Wilkinson Microwave Anisotropy Probe.

FURTHER READING

Allaway, Howard. *The Space Shuttle at Work*. NASA SP-432. Washington, D.C.: Government Printing Office, 1979.

Chant, Christopher. *Space Shuttle*. New York: Exeter Books, 1984.

Godwin, Robert, ed. *Space Shuttle STS Flights 1-5: The NASA Mission Reports*. Burlington, Ont.: Apogee Books, 2001.

Hallion, Richard. *On the Frontier: Flight Research at Dryden, 1946-1981*. Washington, D.C.: Government Printing Office, 1984.

Harland, David M. *The Space Shuttle: Roles, Missions, and Accomplishments*. New York: John Wiley, 1998.

Harrington, Philip S. *The Space Shuttle: A Photographic History*. San Francisco, Calif.: Brown Trout, 2003.

Jenkins, Dennis R. *Space Shuttle: The History of the National Space Transportation System—The First 100 Missions*. Stillwater, Minn.: Voyageur Press, 2001.

Kerrod, Robin. *Space Shuttle*. New York: Gallery Books, 1984.

Lewis, Richard S. *The Voyages of Columbia: The First True Spaceship*. New York: Columbia University Press, 1984.

Reichhardt, Tony. *Proving the Space Transportation System: The Orbital Flight Test Program*. NASA NF-137-83. Washington, D.C.: Government Printing Office, 1983.

Slayton, Donald K., with Michael Cassutt. *Deke! U.S. Manned Space: From Mercury to the Shuttle*. New York: Forge, 1995.

Trento, Joseph J. *Prescription for Disaster: From the Glory of Apollo to the Betrayal of the Shuttle*. New York: Crown, 1987.

Wilson, Andrew. *Space Shuttle Story*. New York: Crescent Books, 1986.

—*Russell R. Tobias*

SPECTROSCOPY

THE SCIENCE: Joseph Fraunhofer discovered that sunlight, when passed through a glass prism or a grating, produced a spectrum of colors that contained numerous dark lines. Later investigators showed that these lines were due to specific chemical elements.

THE SCIENTISTS:

Joseph Fraunhofer (1787-1826), skilled glass maker, inventor of the spectroscope, and discoverer of numerous dark lines in the spectrum of sunlight

Gustave Kirchhoff (1824-1887) and

Robert Bunsen (1811-1899), German scientists who used the spectroscope to show that each chemical element emits a unique pattern of spectral lines

Thomas Young (1773-1829), British scientist who first determined the wavelength of light waves by passing them through two narrowly spaced slits

William H. Wollaston (1766-1818), British scientist who observed seven dark lines in the solar spectrum shortly before Fraunhofer's independent discovery

MISSING WAVELENGTHS

In 1704, the renowned British physicist Sir Isaac Newton published a book entitled *Opticks* (*Optics*, 1706). In it he described his wide-ranging investigations into the properties of light. He measured the angular dispersion of sunlight into a spectrum of colors by using a triangular glass prism. He also gave a mathematical explanation for the creation of the rainbow from the refraction of sunlight by water droplets in the atmosphere.

About a hundred years later, the British scientist William H. Wollaston saw something in the spectrum of sunlight that neither Newton nor anyone else had noted before. He was using a narrow slit for the sunlight to enter a dark room, where it struck a prism. The resulting spectrum was observed from ten feet away. At that distance, the colors from red to violet were greatly spread out. Wollaston noticed that the continuous spectrum of the Sun had some narrow, dark lines in it. Whereas an ordinary light source viewed through a prism emits a truly continuous spectrum of colors, sunlight appears to have some missing wavelengths. He reported finding seven dark lines but had no explanation for what caused them.

FRAUNHOFER'S SPECTROSCOPE

Twelve years after Wollaston's discovery, Joseph Fraunhofer indepen-dently rediscovered the dark lines in the spectrum of the Sun. He devised a special apparatus, the spectroscope, that enabled him to catalog more than five hundred dark lines, now called Fraunhofer lines in his honor. The spectroscope had a lens that could be pointed at the Sun or any other source of light, followed by a narrow slit. The incoming light beam struck a prism made of flint glass that produced a relatively large angular separa-tion of colors. The spectrum was viewed through an eyepiece attached to a platform that could be rotated, allowing the angle of view to be measured with high precision. The most prominent dark lines were given letter names. Fraunhofer noted that the so-called D line in the solar spectrum exactly matched the angle of sodium light that had been observed previously. However, he was not able to interpret the significance of this observation.

Thomas Young, another British scientist, earlier had shown that a light

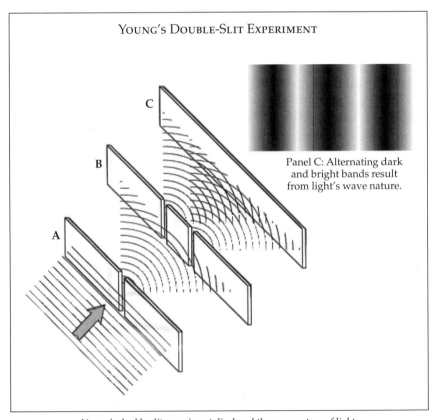

YOUNG'S DOUBLE-SLIT EXPERIMENT

Panel C: Alternating dark and bright bands result from light's wave nature.

Young's double-slit experiment displayed the wave nature of light.

beam, when passed through two slits that are very close together, produced an interference pattern of bright and dark images on a screen. He explained this pattern using the wave theory of light: when two waves are in step, their amplitudes will add to produce a brightness, but when they are half a wavelength out of step, their amplitudes will cancel to produce darkness. Young developed a mathematical formula that used the distance between the two slits and the angles of maximum brightness to calculate the wavelength of the light. Fraunhofer improved on Young's double slit by making a grating, consisting of a large number of closely spaced parallel slits. He wound a thin metal wire back and forth between two threaded screws. By advancing from one thread to the next one, he obtained a closely spaced mesh of wires.

Fraunhofer replaced the prism in his spectroscope with such a grating. The angular separation of colors, or dispersion, was not much better than his result had been with the prism. To improve his observations further, he needed to make the slits in the grating even closer together. As part owner of a glassworks company, Fraunhofer had access to a machine shop. A new grating was made by scribing hundreds of evenly spaced parallel lines on a piece of glass. Light came through the spaces to form a spectrum with high dispersion. With this device, he was able to measure the wavelength of yellow light from a sodium flame with a precision that agrees within one percent of the modern accepted value.

Fraunhofer was not an academic scientist. He was skilled in making glass lenses for optical instruments. He used the solar dark lines as fixed calibration points to measure how the index of refraction of glass varied throughout the spectrum. He learned how to combine lenses of different glass composition into an achromatic system that gave the sharpest possible images. He became famous throughout Europe as the premier supplier of lenses for large telescopes.

EMISSION SPECTRA

Gustave Kirchhoff, a physicist, and Robert Bunsen, a chemist (of Bunsen burner fame), were colleagues at the University of Heidelberg in Germany. In the 1850's, they were studying the spectra of flames that contained various chemicals, such as sodium, potassium, and copper salts. Using a grating in a spectroscope, they observed that each element had a unique spectrum of bright lines. These emission spectra provided them with an unambiguous identification, like a fingerprint, for each element.

Kirchhoff and Bunsen were aware of Fraunhofer's work, thirty-five years earlier, on dark lines in the spectrum of sunlight. In trying to under-

Chemical Fingerprints

In 1860, Gustave Kirchhoff, a physicist, and Robert Bunsen, the chemist for whom the Bunsen burner is named, described the characteristic spectra of some specific elements and speculated on the wide variety of applications of this knowledge:

It is known that several substances have the property of producing certain bright lines when brought into the flame. A method of qualitative analysis can be based on these lines, whereby the field of chemical reactions is greatly widened and hitherto inaccessible problems are solved. . . . The lines show up the more distinctly the higher the temperature and the lower the luminescence of the flame itself. The gas burner described by one of us [Bunsen] has a flame of very high temperature and little luminescence and is, therefore, particularly suitable for experiments on the bright lines that are characteristic for these substances. . . .

In this time-consuming, extensive research, which need not be presented here in detail, it came out that the variety of the compounds in which the metals were used, the differences in the chemical processes of the flames, and the great difference between their temperatures had no influence on the position of the spectral lines corresponding to the individual metals. . . .

Kirchhoff and Bunsen went on to describe their spectroscopic experiments and their outcomes with sodium, lithium, potassium, strontium, calcium, and barium, as well as their conclusions:

Spectrum analysis should become important for the discovery of hitherto unknown elements. If there should be substances that are so sparingly distributed in nature that our present means of analysis fail for their recognition and separation, then we might hope to recognize and to determine many such substances in quantities not reached by our usual means, by the simple observation of their flame spectra. We have had occasion already to convince ourselves that there are such now unknown elements. . . .

Spectrum analysis, which, as we hope we have shown, offers a wonderfully simple means for discovering the smallest traces of certain elements in terrestrial substances, also opens to chemical research a hitherto completely closed region extending far beyond the limits of the Earth and even of the solar system. Since in this analytical method it is sufficient to see the glowing gas to be analyzed, it can easily be applied to the atmosphere of the Sun and the bright stars.

Source: Gustav Kirchhoff and Robert Bunsen. "Chemical Analysis by Observation of Spectra." *Annalen der Physik und der Chemie* (Poggendorff) 110 (1860): 161-189.

stand these lines, Kirchhoff created a crucial experiment. Using a laboratory lamp, he showed that it had a true continuous spectrum with no dark lines. Then he placed a sodium flame between the lamp and the grating. This time the continuous spectrum had a dark line in the yellow region, just at the known wavelength of sodium. Evidently, sodium vapor was absorbing its particular wavelength out of the continuous spectrum.

Kirchhoff and Bunsen proposed the idea that atoms of the chemical elements have distinct "absorption spectra" that match their emission spectra. They were able to show that three prominent Fraunhofer dark lines in the solar spectrum exactly matched the emission wavelengths of potassium. They came to the conclusion that light from the surface of the Sun was being absorbed at fixed wavelengths by sodium, potassium, and other atoms in the Sun's outer atmosphere.

IMPACT

The Fraunhofer dark lines have led to some interesting results. Sir John Lockyer, a British astronomer, in 1868 speculated that a prominent dark line in the solar spectrum, which did not match any element known on earth, might be due to a new element found only on the Sun. He named it helium, after the Greek word for the Sun. Some thirty years later, helium gas eventually was found deep in mine shafts. Helium became a valuable resource for various technological applications, including lighter-than-air balloons.

Fraunhofer dark lines are found in the spectra not only of the Sun but also of all stars. Astronomers can use a telescope to focus on one star at a time and can record its spectrum on photographic film. In some cases, the Fraunhofer lines show a shift toward a longer wavelength—that is, toward the red end of the spectrum. Such a "redshift" comes about when a star is moving away from the Earth at high speed. This phenomenon is like the drop in frequency that one hears when an ambulance with a siren is traveling away from the listener. The redshift in the Fraunhofer lines from distant stars is the primary evidence for an expanding universe.

Spectroscopy was extended to other parts of the electromagnetic spectrum as new instrumentation became available. For example, infrared spectra are the primary means to obtain information about the structure of molecules. Gamma-ray spectroscopy has become a highly developed method of analysis that can detect impurities in materials as small as a few parts per billion. Fraunhofer's spectroscope was the starting point for many practical applications in analytical chemistry, astronomy, medical research, and other technologies.

See also Buckminsterfullerene; Expanding Universe; Galaxies; Gamma-Ray Bursts; Inflationary Model of the Universe; Isotopes; Lasers; Optics; Radioactive Elements; Schrödinger's Wave Equation; Speed of Light; Superconductivity; Superconductivity at High Temperatures; Thermodynamics: First and Second Laws; Thermodynamics: Third Law; Water; Wave-Particle Duality of Light; X Radiation; X-Ray Crystallography; X-Ray Fluorescence.

FURTHER READING

Connes, Pierre. "How Light Is Analyzed." *Scientific American* (September, 1968): 72-82.
Jackson, Myles. *Spectrum of Belief: Joseph Fraunhofer and the Craft of Precision Optics*. Cambridge, Mass.: MIT Press, 2000.
Walker, James S. *Physics*. 2d ed. Upper Saddle River, N.J.: Pearson/ Prentice Hall, 2004.

—*Hans G. Graetzer*

SPEED OF LIGHT

THE SCIENCE: In 1676, Ole Rømer's measurement of the speed of light was the first clear demonstration that light travels with a finite velocity. Although his value was off by about 25 percent, his method was correct in principle. A more accurate value was obtained fifty years later by James Bradley using another astronomical method.

THE SCIENTISTS:
Ole Rømer (1644-1710), Danish astronomer whose study of Jupiter's moons helped determine the speed of light
Galileo Galilei (1564-1642), Italian physicist and astronomer who made one of the first attempts to measure the speed of light
Christiaan Huygens (1629-1695), Dutch physicist and astronomer who assisted Rømer in calculating the speed of light
James Bradley (1693-1762), third Astronomer Royal in England, whose measurements of the aberration of starlight led to the first accurate measurement of the speed of light
Gian Domenico Cassini (1625-1712), Italian astronomer and first director of the Paris Observatory

BEYOND INFINITY

Before the seventeenth century, scientists believed that the speed of light was infinite. In about 1607, Galileo attempted to measure the speed of light with the aid of an assistant on a hilltop at some distance away with a covered lamp. When the assistant saw Galileo uncover a similar lamp, he then uncovered his lamp and Galileo tried to observe the time for the light to travel to the assistant and back again. He concluded that the speed of light was either instantaneous or extremely rapid.

The first observations showing that the speed of light is finite were made by the Danish astronomer Ole Rømer in Paris in 1675. Using the new pendulum clock invented in 1657 by Christiaan Huygens, a fellow foreign member of the Royal Academy of Sciences, Rømer determined that the 42.5-hour period of Jupiter's moon Io had an orbital period that was a maximum of 13 seconds longer when the Earth was moving away from Jupiter and 13 seconds less time when it was approaching (42.5 hr. ± 13 seconds). He recognized that this phenomenon occurred because the light took longer to reach the Earth as it moved away from Jupiter and shorter as the Earth moved toward Jupiter in each 42.5-hour orbit of Io.

RØMER'S CALCULATIONS

To determine the range of variations in the orbital period, Rømer observed consecutive eclipses of Io as it passed behind Jupiter, noting the times when it emerged from each eclipse. Since these emergences of the Moon from eclipses were not instantaneous events, there were some errors in his measurements. From these variations, he calculated that light would take about 22 minutes to cross Earth's orbit (compared with a modern value of about 16 minutes). On November 22, 1675, Rømer read a paper to the science academy, in which he announced that an eclipse of Jupiter's moon Io would occur about 10 minutes later than the time predicted from the average orbital period as measured in 1668 by Gian Domenico Cassini, director of the Paris Observatory and also a foreign member of the science academy.

Working with the aid of Huygens, Rømer combined the 22-minute time for light to cross Earth's orbit with the diameter of Earth's orbit as determined by Cassini in 1671, a value that was 7 percent too small. By taking the ratio of the distance to the time, he found the speed of light to be about 230 million meters per second, or about three-fourths of the modern value of nearly 300 million meters per second.

Rømer published his discovery in a short paper entitled "Demonstration touchant le mouvement de la lumière trouvé" (demonstration concerning the discovery of the movement of light) in the *Journal des Savants* on December 7, 1676. At the request of the Danish king, Rømer returned to Denmark in 1681 as royal mathematician and professor of astronomy at Copenhagen University.

BRADLEY'S CALCULATIONS

The first accurate measurement of the speed of light was made some fifty years after Rømer's measurements by the English astronomer James Bradley in 1728, also using an astronomical method. Bradley was trying to find evidence for the Earth's motion around the Sun by measuring the annual stellar parallax, the shifting angle of the stars that should result from Earth's motion in a six-month period. Rømer also had attempted to measure this parallax, but he had failed to detect any change. Although Bradley also failed to measure any parallax, he did notice a relatively large shift in angle of one second of arc in just three days and in the wrong direction to qualify as the annual parallax.

According to some accounts, Bradley's explanation of the anomalous star angles he observed occurred to him while sailing on the Thames River and noticing how a steady wind caused the wind vane on the mast to shift relative to the boat as it changed directions. He reasoned that the apparent shift in star angles resulted from the orbital motion of the Earth relative to the constant speed of light. This "aberration of starlight" is similar to the apparent angle of vertically falling raindrops relative to a moving observer. The angle of stellar aberration is given approximately by the ratio of the Earth's forward orbital speed to the speed of light. Careful measurements of this angle combined with the known speed of the Earth allowed Bradley to obtain a value of 295 million meters per second for the speed of light, slightly too small (but by less than 2 percent).

Bradley's precise measurements of stellar aberration not only improved the value for the finite speed of light but also provided the first direct evidence for the motion of the Earth as suggested by the Copernican theory some two hundred years earlier. Further careful measurements of star angles by Bradley revealed in 1732 the nodding motion of the Earth's axis, called nutation, resulting from variations in the direction of the gravitational pull of the Moon. For these achievements, he was named the third Astronomer Royal in England. His value for the speed of light was not corrected until terrestrial measurements were begun in mid-nineteenth century France, when the original method of Galileo was improved by using reflected light and rapid timing by rotating wheels.

Impact

Even though Ole Rømer's value for the speed of light was about one-quarter too small, his method was correct and revealed that light has a finite speed. By showing that light travels nearly one million times faster than sound, Rømer provided evidence that eventually showed that light cannot consist of a mechanical propagation like sound, but is actually an electromagnetic wave as demonstrated in the nineteenth century.

OLE RØMER: HEAT AND LIGHT

Ole Rømer was born in Åarhus, the largest city in Jutland, Denmark, on September 25, 1644, and studied astronomy in Copenhagen. He assisted in determining the exact location of Tycho Brahe's observatory on the island of Hveen in 1671 and then went to Paris in 1672. He remained for nine years at the new Paris Observatory of the Royal Academy of Sciences, making careful observations of the moons of Jupiter. In 1675, he discovered an inequality in the motion of the moon Io, the closest and fastest of the four large moons of Jupiter discovered by Galileo in 1610.

In addition to calculating the speed of light, Rømer played a key role in the development of the modern thermometer, paving the way for Daniel Fahrenheit's work. Rømer was particularly interested in creating a reproducible thermometer, so that experiments and observations from widely differing locales could be compared. Due to problems with hand-blowing the hollow glass tubes that were used in making thermometers, it was impossible to make them physically identical. As a result, it was necessary to find some other way to determine when they all indicated the same temperatures. Rømer's solution was to calibrate each thermometer against known reference points (such as the melting point of ice and the boiling point of water) so that it would be possible to have all the thermometers measuring temperature equally even if they were not structurally identical. It remained to assign numerical values to the various points on his scale. Rømer experimented with a number of different scales, setting various numbers for the reference points. Rømer still had not settled upon a workable scale when Daniel Fahrenheit arrived to discuss questions of measurement with him.

Historians of science would subsequently argue intensely about the extent of Rømer's role in inspiring Fahrenheit's work in thermometers and temperature scales, until the discovery of a letter in an archive in Leningrad (St. Petersburg, Russia). In the letter, Fahrenheit recounts experiments that he and Rømer performed together, which led him to an interest in improving the mechanism of both thermometers and barometers.

Bradley's improved method for measuring the speed of light began a quest for precision that finally revealed the true nature of light and gave the first direct evidence for the motion of the Earth. Terrestrial measurements a century after his work gave the most accurate values for the speed of light and revealed that light travels more slowly in water than in air, confirming the wave nature of light. When electromagnetic studies showed that light is propagated by electric and magnetic fields, the speed of the resulting electromagnetic waves could be calculated from electric and magnetic constants as measured in the laboratory, and the result matched the observed speed of light. In Albert Einstein's theory of relativity, the speed of light is seen as one of the fundamental constants of the universe.

See also Cepheid Variables; Diffraction; Gravitation: Einstein; Inflationary Model of the Universe; Lasers; Optics; Photoelectric Effect; Quantized Hall Effect; String Theory; Superconductivity at High Temperatures.

FURTHER READING

Alioto, Anthony M. *A History of Western Science*. 2d ed. Upper Saddle River, N.J.: Prentice Hall, 1993.
Cohen, I. Bernard. *Roemer and the First Determination of the Velocity of Light*. New York: Burndy Library, 1944.
Crump, Thomas. *A Brief History of Science as Seen Through the Development of Scientific Instruments*. New York: Carroll & Graf, 2001.
Huygens, Christiaan. *Treatise on Light*. Vol. 34 in *Great Books of the Western World*, edited by R. M. Hutchins. Chicago: William Benton, 1952.
—*Joseph L. Spradley*

SPLIT-BRAIN EXPERIMENTS

THE SCIENCE: The fact that the two halves of the human brain can function separately, and the ways they interact, were demonstrated by Roger W. Sperry and his colleagues in a series of brilliant experiments.

THE SCIENTISTS:
Roger W. Sperry (1913-1994), neurophysiologist who won the 1981 Nobel Prize in Physiology or Medicine
Michael S. Gazzaniga (b. 1939), graduate student
Ronald Myers (b. 1929), graduate student

Splitting the Brain

The nature of the human body is such that many of its parts come in two halves. The halves of these paired organs—such as lungs, kidneys, and eyes—generally perform similar or identical functions. It was long assumed that the two halves of the brain likewise have a single function. Scientists were therefore surprised to discover not only that the two halves of the brain perform different activities but also that, in certain cases, each half can function independently.

In the early 1950's, Roger W. Sperry began to research the function of the corpus callosum. This narrow bundle of nerve cells, containing some 200 million neurons, connects the two halves of the brain. At the University of Chicago, with a graduate student, Ronald Myers, Sperry severed the corpus callosum in a cat. He covered up the cat's right eye, forcing it to see only through its left eye, and taught the cat a simple task. Things seen by the left eye are stored in the right half of the brain. By forcing the cat to use its left eye, Sperry was making sure that all learning would be stored in the right half only. Yet when its left eye was closed and right eye open (which forced the cat to use the left half of the brain and any learning stored there),

it was unable to perform the same task. Sperry concluded that the information absorbed through the left eye could not pass between the two halves when the corpus callosum was cut. The right half could, however, learn the task all by itself. It was as if the cat had two separate brains, each of which could function independently when separated from the other.

There was reason to doubt that these findings would be relevant to human brains. The corpus callosum had been cut in a number of humans as a last resort in controlling severe epilepsy. Doctors reasoned that if the connection between the two halves was

Roger W. Sperry. (The Nobel Foundation)

cut, an epileptic seizure occurring in one half of the brain would leave the other half unaffected. The surgery not only proved to be effective in limiting the spread of epileptic seizures but also, for reasons still unknown, actually decreased the frequency of such seizures. Fortunately for the patients, there were no obvious changes in personality, intelligence, or mental functioning. This, however, suggested to scientists that the corpus callosum had no important function in humans.

RIGHT BRAIN, LEFT BRAIN

More careful behavioral studies were begun by Michael S. Gazzaniga, a graduate student, in Sperry's laboratory. His first subject was a forty-eight-year-old war veteran whose corpus callosum had been severed to control his epilepsy. The experimental procedure was simple: A picture of some everyday object was flashed in front of one eye or the other, and the subject was asked to report what he saw. A normal person would report having seen an object no matter which eye was involved, but Gazzaniga's subject reported seeing only objects viewed by his right eye (and perceived by the left hemisphere). When a picture was flashed before his left eye (and perceived by the right hemisphere), he denied having seen anything. The right hemisphere was not "blind," it simply could not "speak": When the subject was asked to point to an object he had seen, he was able to do so, indicating that his right hemisphere had, in fact, perceived the object.

Sperry and Gazzaniga thus solved the problem of the elusive function of the corpus callosum. When a normal person sees an object in the left visual field, the right hemisphere, which obtained the information, sends it through the corpus callosum to the left hemisphere, which can then verbalize a response about what the individual saw. The corpus callosum thus allows communication between the two hemispheres. For the subject in the experiment, this had been impossible because of the severed connection.

Although the right hemisphere was initially considered inferior because it lacked the verbal ability of the left hemisphere, subsequent research showed that the right hemisphere can understand the vocabulary of a ten-year-old. Although unable to direct the mouth to speak, the right hemisphere could direct the left hand to move plastic letters so as to spell out the answers to certain questions.

Gazzaniga tested split-brain patients in another study, in which they were required to arrange a set of blocks to match a design in a picture. The left hand (guided by the right half of the brain) was superior at this task. The scientists concluded that the right hemisphere is important for spatial skills. In related work, Doreen Kimura, studying normal individuals, showed that the left hemisphere is better at interpreting verbal information, while the right hemisphere is better at identifying melodies.

Impact

The acceptance of the idea of hemispheric specialization had some positive effects in the general population, since people gained a better understanding of nonverbal forms of intelligence. Intelligence tests, which had traditionally measured what have come to be considered left-brain activities, are placing more emphasis on measuring and appreciating types of intelligence that operate in the right brain.

Sperry's studies also had some impact on the understanding of certain disorders. Dyslexia, for example, is a disorder that makes it difficult to learn to read. Children who suffer from dyslexia tend to show less than the usual right-hemisphere specialization for spatial relations. The idea that reading disorders are biologically based and not the result of the children's misbehavior has led to more flexibility in dealing with these disorders.

Split-brain research raises certain philosophical considerations: Is the split brain also a split mind? Are there two separate consciousnesses in a single individual? Controversy about the meaning of split-brain research continues long after the research was first announced in the 1960's. For stimulating this controversy, Sperry was awarded the Nobel Prize in Physiology or Medicine in 1981.

See also Manic Depression; Pavlovian Reinforcement; Psychoanalysis; REM Sleep.

Further Reading

Baskin, Yvonne. "Emergence: Roger Sperry." In *The Omni Interviews*, edited by Pamela Weintraub. New York: Ticknor & Fields, 1984.

Gazzaniga, Michael S. *The Bisected Brain*. New York: Appleton-Century-Crofts, 1970.

_____. *The Social Brain: Discovering the Networks of the Mind*. New York: Basic Books, 1985.

Segalowitz, Sid J. *Two Sides of the Brain: Brain Lateralization Explored*. Englewood Cliffs, N.J.: Prentice-Hall, 1983.

Sperry, Roger. "Some Effects of Disconnecting the Cerebral Hemispheres." *Science* 217 (1982): 1223-1226.

—*Judith R. Gibber*

SPONTANEOUS GENERATION

THE SCIENCE: Lazzaro Spallanzani was among the first to show experimentally that living organisms—such as maggots in rotting meat—could not simply appear out of nowhere. Though his work was not considered conclusive on the subject, it represented the beginnings of a modern view of biology.

THE SCIENTISTS:
Lazzaro Spallanzani (1729-1799), Italian biologist
Francesco Redi (1626-1697), Italian physician
Georges-Louis Leclerc, Comte de Buffon (1707-1788), French naturalist
     and author of the first comprehensive natural history
John Tuberville Needham (1713-1781), Catholic priest and collaborator
     with Buffon

LIFE FROM NONLIFE?

Naturalists before the eighteenth century had observed many instances of what seemed to them to be the "spontaneous generation" of life. Meat left out would sprout maggots, and frogs could similarly emerge from apparently simple mud. There was no mystery associated with these seeming miracles, however. Indeed, spontaneous generation made perfect sense to those, now called "vitalists," who believed in a Creator who had the ability to produce life from abiotic matter. If life had first originated in this manner, the reasoning went, life could appear again through similar means. There were also nonreligious explanations for the phenomenon. The ancient Greek philosopher Aristotle, during the fourth century B.C.E., believed that humidity provided a form of life force to dry objects. Later naturalists argued that mud could produce frogs or eels and even had recipes for the formation of life.

The first significant experiments to address the subject of spontaneous generation were carried out by Franceso Redi in 1668. An Italian physician and member of del Cimento Academy, Redi designed a series of experiments in which putrefying meat was placed in vessels. Some vessels were covered with gauze or were completely sealed, while others served as uncovered controls. Redi observed that only meat that was accessible to flies developed maggots. The French physicist René-Antoine Ferchault de Réaumur, more famous for development of an alcohol thermometer, would later directly observe flies depositing eggs in food.

The debate over spontaneous generation continued for more than a cen-

Lazzaro Spallanzani. (Library of Congress)

tury, and the development of the microscope, resulting in the discovery of microscopic "animalcules" by Antoni van Leeuwenhoek, only added to the debate. Leeuwenhoek's work established that an entire world of living things existed beyond the ability of the human eye to see. Even if it were established that the organisms of the visible world were incapable of spontaneous generation, therefore, it might still be the case that microscopic animalcules could appear spontaneously. Support for this view could be found in experiments carried out by the British clergyman John Tuberville Needham in 1745. Since it was known by then that heat could kill microorganisms, Needham boiled chicken broth and placed it in sealed vessels. Despite this treatment, microorganisms would still appear in the broth.

Needham later went to Paris, where he met and began a collaboration with the comte de Buffon. Buffon was well noted for his contributions to the growing field of comparative anatomy in the massive work *Histoire naturelle, générale et particulière* (1749-1789; *Natural History, General and Particular*, 1781-1812). Though many of Buffon's views on the similarities of species and the age of the Earth were still controversial at the time of their publication, he was well enough respected that his support for Needham's views lent credibility to the arguments in favor of spontaneous generation.

HEATING THE DEBATE

Lazzaro Spallanzani had a differing interpretation of Needham's results, however: He believed that the broth had been contaminated before being sealed. His criticism of Needham's techniques formed the basis for a 1765 dissertation on the subject. He also, more importantly, devised a set of practical tests to confirm his hypothesis that Needham's samples must have been contaminated. Spallanzani's experimental procedure was relatively simple: He boiled his samples for varying periods to ensure that nothing survived and then sealed the mixture in an airtight container. Beginning with a duration of forty-five minutes, Spallanzani tested various boiling periods, observing whether anything would still grow in the broth after each test.

Spallanzani determined that extensive boiling prevented microorganisms from growing, resulting in the medium remaining sterile. To refute the potential counter-argument that boiling destroyed a "life force" that had existed in the broth and that was necessary for spontaneous generation to occur, Spallanzani sealed his vessels with semipermeable barriers. He created seals with pores of various sizes, and he observed that the number of organisms that returned to the boiled broth was a function of the pore size. The implication of this result was that contamination from air had been the source of growth in Needham's experiments.

Spallanzani also observed that there existed several classes of organisms that differed in their sensitivity to heat. One class, probably protozoa, was highly sensitive to heat, and Spallanzani labeled this class "superior animalcula." On the other hand, the class he named "lower class animalcula," probably bacteria, was less sensitive. Thus, a nonrigorous application of heat in an experiment might kill only the more sensitive microscopic organisms, leaving the less sensitive ones to "appear spontaneously" afterward.

Despite these results, Spallanzani's experiments did not resolve the issue of spontaneous generation in the minds of all scientists. The experiments' results admitted of different interpretations, especially after Joseph Priestley discovered oxygen in 1774. When it was discovered that oxygen itself was driven from Spallanzani's experimental vessels during the heating process, some scientists argued that this newly discovered gas was necessary to activate the "vital force" that caused life to appear from nothing. It would remain for Louis Pasteur in the 1860's to resolve the argument to the satisfaction of the entire scientific community. After all, spontaneous generation had been believed to exist, in one form or another, for at least two thousand years. Such an entrenched belief could not be eliminated with anything less than utterly conclusive proof.

The theory explained quite efficiently an otherwise mysterious phenomenon that, in the days before preservatives and refrigeration, was extremely common—the sudden appearance of maggots, flies, or other biological contaminants on seemingly clean foods. The advances in experimental design in the sciences during the sixteenth and seventeenth centuries allowed naturalists to test the theory. Among the first to do so was Francesco Redi. However, despite Redi's initial results, which seemed to demonstrate that infestation of maggots in meat resulted from flies, not decay, the belief in spontaneous generation continued for years. Even Redi himself was not completely convinced that spontaneous generation was impossible.

IMPACT

Spallanzani's work, following from Redi's, was carried out during a period in the eighteenth century in which science was undergoing significant advancement. The development of the microscope in the previous century had allowed the observation of the very small. The significance of such observations when applied to germs or contamination, however, was still misunderstood. Nevertheless, Spallanzani continued the earlier experiments of Redi on the subject by demonstrating that if water was sterilized in a sealed container, spontaneous formation of life could not occur.

Where Redi's experiments, important in their use of biological controls, had demonstrated that relatively large forms of life would not spontaneously appear, Spallanzani showed that even microscopic organisms could be eliminated through the use of heat. In the process, he brought an experimental approach to the study of such organisms, complementing the scientific approach to the macroscopic world developed by the comte de Buffon. Nicolas Appert, a French cook, several decades later would apply Spallanzani's approach in developing a method of canning as a way to preserve food.

In the immediate wake of Spallanzani's experiments, however, the debate concerning spontaneous generation was not yet settled. Criticism of Spallanzani's experiments centered on the lack of a formal, scientifically rigorous understanding of the nature of life—a lack not fully corrected even today. Various vitalist arguments about a "life force" that made generation possible complicated matters tremendously: If this hypothetical force was itself intangible, it would be difficult or impossible to prove that it had not been altered by heat, the lack of air, or other effects of Spallanzani's experiments.

Moreover, as a result of the state of experimental design in the 1760's, Spallanzani himself never came to the conclusion that microorganisms were present in the air. The fact that there was a gas called oxygen in the air, indeed the fact that the air was composed of a mixture of different gases, was new information added to the debate in the 1770's. As a result, one could still not rule out the possibility that it was the presence of oxygen that was required for spontaneous generation to occur.

It would be nearly one hundred years before Louis Pasteur ended the debate with what became known as the "swan-neck" flask experiment. In this experiment, Pasteur placed the experimental sample in a flask that remained unsealed, but he created a curved neck for the flask that allowed air to enter normally but kept microorganisms out; the solution under such circumstances remained sterile. Still, Spallanzani's work remained an important step in the evolution of biological knowledge.

See also Cell Theory; Contagion; Fossils; Germ Theory; Lamarckian Evolution; Microscopic Life; Photosynthesis.

FURTHER READING

Brock, Thomas. *Milestones in Microbiology, 1546-1940*. Washington, D.C.: ASM Press, 1999.
Buffon, Georges-Louis Leclerc. *Natural History: General and Particular*. Translated by William Smellie. Bristol, Avon, England: Thoemmes Press, 2001.
Geison, Gerald. *The Private Science of Louis Pasteur*. Princeton, N.J.: Princeton University Press, 1995.
Lechevalier, Hubert, and Morris Solotorovsky. *Three Centuries of Microbiology*. New York: Dover, 1974.
Lennox, J. "Teleology, Chance, and Aristotle's Theory of Spontaneous Generation." *Journal of History of Philosophy* 19 (1981): 219-238.
Strick, James. *Sparks of Life: Darwinism and the Victorian Debates over Spontaneous Generation*. Cambridge, Mass.: Harvard University Press, 2000.
—*Richard Adler*

STELLAR EVOLUTION

THE SCIENCE: Henry Norris Russell used the color-luminosity relationship of stars to work out a theory of how stars change over time.

THE SCIENTISTS:
Henry Norris Russell (1877-1957), American astronomer who discovered the color-luminosity relationship and codeveloped the Hertzsprung-Russell diagram that led to his theory of stellar evolution
Ejnar Hertzsprung (1873-1967), Danish astronomer and photographer, who established that there is a relationship between a star's color and its luminosity
Sir Joseph Norman Lockyer (1836-1920), British astronomer who developed a theory of stellar evolution later elaborated by Russell

A Starry Garden

Sir William Herschel described the starry sky as a garden, wherein one sees stars in varying stages of their lives, as in a garden or a forest one sees plants and trees in varying stages of early growth, maturity, and death. The assumption of seeing stars of different ages and that stars change as they age was an important prerequisite for the formation of theories of stellar evolution. The advent of increasingly sophisticated techniques for classifying stars in the late nineteenth and early twentieth centuries brought a wealth of data on spectral types from which a theory of stellar evolution could be built.

Classification Systems

Sir Joseph Norman Lockyer, working in the late 1800's, used the simple classification systems of Angelo Secchi and Hermann Karl Vogel—which placed stars in one of four categories—to develop a scheme for stellar evolution. This scheme was based on current theory regarding the energy source for stars and the physical forces shaping their life histories. At the time, physicists believed that a star's radiation was heat and light, which was released as the star contracted under the force of gravity. A star was believed to form when enough interstellar matter accumulates in one place to begin to exert gravitational attraction on itself and to form a sphere. The star then begins to contract inward under the force of gravity and to heat up and to shine. Eventually, the collapse is halted when a critical density is reached, and the star begins to cool off and die. Lockyer used the spectral classes of the time to identify a sequence of stages through which it was believed that all stars pass.

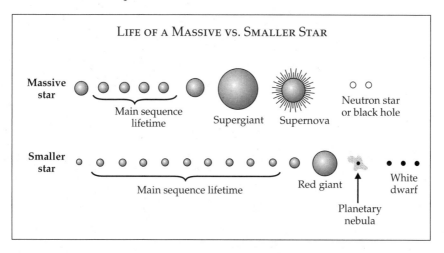

LIFE OF A MASSIVE VS. SMALLER STAR

Henry Norris Russell had at his disposal a more sophisticated system of classification, involving seven classes of stars. Also, many more stars had been classified while Russell was conducting his research. This was largely the result of a program carried out at Harvard College Observatory under the direction of Edward Charles Pickering at the beginning of the twentieth century, in which stars were given classifications based on their spectra. A star's spectrum, or the bands of color and darkness produced when its light is spread out by a prism or grating, contains dark lines that can be used to classify stars. The researchers at Harvard College Observatory looked at thousands of such spectra and classified their associated stars, producing massive catalogs of information on stellar types. Russell was able to use this information in developing his scheme of stellar evolution.

The Hertzsprung-Russell Diagram

Because stars cannot be directly examined in the laboratory, astronomers are forced to deduce their characteristics from things that can be ob-

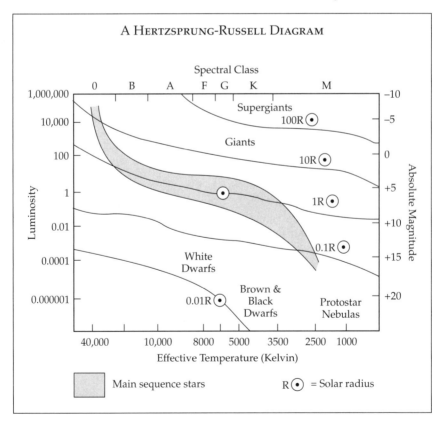

served, such as their spectral type or their brightness. Russell was faced with the question of how to identify a star's characteristics and thus its stage of evolution by its visible characteristics. A star's spectral type was almost universally believed at the time to be linked to its surface temperature and its color, and the different types were the result of differing temperatures; however, no consensus had been reached on the cause of differences in brightness. Russell showed that differences in brightness were a result of variations in density. Thus, spectral type was related to surface temperature and color, and brightness was related to density.

Russell plotted data on spectral types versus data on the absolute brightness (luminosity) of stars (that is, a star's true brightness, after its brightness as seen from Earth is corrected for its distance). In 1913, he produced a plot of spectral type versus brightness. Ejnar Hertzsprung had made a similar diagram in 1911. This type of plot is known today as a Hertzsprung-Russell, or H-R, diagram. Russell then used this plot to view the relationship between brightness (and density) and spectral type (and color and temperature).

Russell presented his diagram to the Royal Astronomical Society on June 13, 1913, and to the American Astronomical Society in Atlanta, Georgia, on December 30, 1913. He also offered his interpretation of the diagram, in terms of stellar evolution.

Main and Giant Sequence Stars

Most stars fell either on a diagonal band stretching across the plot (the "main sequence") or on a horizontal strip across the top of the plot (the "giant sequence"). The names given to these two sequences were based on work by Hertzsprung and others which determined that the stars on the giant sequence were much larger than the stars on the main sequence.

On the main sequence, stars vary in brightness and color, with stars ranging from bright blue ones to dim red ones. On the giant sequence, stars have a fairly constant brightness but vary in type (color). These two groups or sequences of stars were explained by Russell in terms of the age of stars in each sequence. He used the idea that a star's evolution is driven by gravity alone, and that a star begins its life as cool, red, dim, and diffuse, and grows increasingly dense, bright, and hot (with an associated color change) as it contracts. Once it has contracted as far as it can so that no more gravitational energy is available to it, it begins to cool off and become less bright and more red.

Russell hypothesized that the large red stars at one end of the giant sequence are the youngest of stars and that they represent the earliest stages in a star's life, when stars are very diffuse and just beginning their gravitational collapse. As a star collapses, it becomes more dense and begins to change color and spectral type as it moves across the giant sequence; it eventually brightens and leaves the giant sequence for the main sequence. At its hottest point, which Russell believed to be the midpoint of its life, the star was at the top of the main sequence among the brightest and bluest stars. As it then began to cool, while continuing to become denser, it slid down the main sequence from being a hot blue star to being a yellow star like the Sun and finally to being a dim red star, very dense and near the end of its life.

Thus, this track, along the giant sequence and down the main sequence, was thought to be a path of increasing density and increasing age. There were two sorts of red stars: young, diffuse, large ones of increasing temperature and old, dense, small ones of decreasing temperature. A red star could therefore be at either end of its lifetime; Hertzsprung had demonstrated earlier that the spectra of the two types of red stars were different, and thus enabled astronomers to tell whether a red star was old or young.

Russell presented a concise and straightforward scheme of stellar evolution, which neatly fit the known data in terms of the accepted explanation for why stars shine and how they form, exist, and die. He was able to use his diagram to illustrate succinctly the life-stages of a star as he hypothesized them. His work on the temperature and density of stars, as related to spectral type and brightness, was confirmed by later work. Although his evolutionary scheme later required major revision, it was still an important step in the understanding of the "garden" of varying stars we see.

IMPACT

Hertzsprung-Russell (H-R) diagrams were used by many astronomers immediately after Russell first presented one in 1913, and they are an important tool in astrophysics today. Walter Baade was able to compare H-R diagrams for groups of stars to show that there are, in fact, two populations of stars (one much older than the other) and that each type has its own distinct H-R diagram. This work had important cosmological implications. The H-R diagrams of clusters of stars in the Milky Way have been studied by Robert Julius Trumpler, Bengt Georg Daniel Strömgren, and Gerard Peter Kuiper, among others, to work out theories of stellar formation and evolution.

The discovery that nuclear fusion powers stars for most of their lifetimes, rather than gravitational collapse, brought about drastic revisions in Russell's scheme. Russell's work was important, however, in that it was an early attempt to deduce, from observable quantities, the life cycles of stars. His use of the diagram was a key step in the developing science of astrophysics. Astronomers' knowledge of the causes of a star's observable properties, as plotted on the diagram, changed as they learned of nuclear fusion and nuclear science. However, the method of using the H-R diagram as a clue to a star's properties and life cycle has remained the same. Russell pioneered a practice that continues today.

In his explanation of how the H-R diagram reveals the evolution of stars, Russell gave at least a hint of what was to be discovered later about nuclear power fueling the stars. He suggested that perhaps there is a type of energy release related to radioactivity, which could counteract the gravitational pull inward for a brief period and give a star a longer lifetime than it would have had otherwise. He thought this would not be an important enough effect to change the overall life cycle of the star as he described it.

Today, however, it is known that while a star starts to form because a cloud of material collapses under the influence of gravity, eventually conditions become hot enough in the center of the forming star that nuclear fusion begins to occur. The star then lives out most of its life cycle in one spot on the main sequence, its gravitational pull inward balanced by the pressure outward resulting from energy being released in nuclear fusion. Gravity becomes important again at the end of the star's lifetime, where its fate is determined by the amount of mass it contains. (This is also the factor that determines how long the star lives and where on the main sequence it appears, that is, its brightness and color.)

Much has been learned about such exotic objects as white dwarfs, neutron stars, and black holes, which are the end products of evolution for various masses of stars. Without the H-R diagram, and the foundation of knowledge it offers for the understanding of the interrelationships among a star's density, brightness, temperature, and spectral type, astronomers could not have arrived at their current understanding.

See also Big Bang; Black Holes; Brahe's Supernova; Cassini-Huygens Mission; Cepheid Variables; Chandrasekhar Limit; Copernican Revolution; Extrasolar Planets; Galactic Superclusters; Galaxies; Hubble Space Telescope; Neutron Stars; Pulsars; Quasars; Radio Astronomy; Radio Galaxies; Radio Maps of the Universe; X-Ray Astronomy.

FURTHER READING

Abell, George O. *Realm of the Universe*. 3d ed. New York: Saunders College Publishing, 1984.

Degani, Meir H. *Astronomy Made Simple*. Rev. ed. Garden City, N.Y.: Doubleday, 1976.

Moore, Patrick. *Patrick Moore's History of Astronomy*. 6th rev. ed. London: Macdonald, 1983.

Pagels, Heinz. *Perfect Symmetry*. New York: Simon & Schuster, 1985.

Pannekoek, A. *A History of Astronomy*. New York: Barnes & Noble Books, 1961.

Rigutti, Mario. *A Hundred Billion Stars*. Translated by Mirella Giacconi. Cambridge, Mass.: MIT Press, 1984.

Struve, Otto, and Velta Zebergs. *Astronomy of the Twentieth Century*. New York: Macmillan, 1962.

Trefil, James S. *Space, Time, Infinity: The Smithsonian Views the Universe*. New York: Pantheon Books, 1985.

Vaucouleurs, Gérard de. *Discovery of the Universe*. London: Faber & Faber, 1957.

—*Mary Hrovat*

STEM CELLS

THE SCIENCE: Stem cells, which can be manipulated to create unlimited amounts of specialized tissue, may be used to treat a variety of diseases and injuries that have destroyed a patient's cells, tissues, or organs. Stem cells could also be used to gain a better understanding of how genetics works in the early stages of cell development and may play a role in the testing and development of drugs.

THE SCIENTISTS:
Ernest Armstrong McCulloch (b. 1926?) and
James Edgar Till (b. 1931?), Canadian cellular biologists at the Ontario Cancer Institute who won the 2005 Lasker Award for their pioneering research in proving the existence of stem cells
James Thomson (b. 1958), developmental biologist at the University of Wisconsin, Madison, who first isolated embryonic stem cells in 1998

TYPES OF STEM CELLS

Stem cells are defined by their ability to renew themselves, their lack of differentiation, and their ability to diversify into other cell types. There are three major classes of stem cells: totipotent, pluripotent, and multipotent. Totipotent cells can differentiate to become all of the cells that make up an embryo, all of the extraembryonic tissues, and all of the postembryonic tissues and organs. Pluripotent cells have the potential to become almost all of the tissues found in an embryo but are not capable of giving rise to supporting cells and tissues. Multipotent cells are specialized stem cells capable of giving rise to one class of cells.

A fertilized egg, or zygote, is totipotent. The zygote first divides into two cells about one day after fertilization and becomes an embryo. The embryonic cells remain totipotent for about four days after fertilization. At that point, the embryo consists of about eight cells. As the cells of the embryo continue to divide, they form a hollow sphere. The approximately fifty to one hundred cells on the inner side of the sphere are pluripotent and will continue developing to form the embryo, while the cells on the outer surface will give rise to the extraembryonic tissues, such as the placenta and the umbilical cord.

Multipotent stem cells are found in a variety of tissues in adult mammals and are sometimes referred to as adult stem cells. They are specialized stem cells that are committed to giving rise to cells that have a particular function. Identities of some multipotent stem cells have been confirmed. Hematopoietic stem cells give rise to all the types of blood cells. Mesenchymal stem cells in the bone marrow give rise to a variety of cell types: bone cells, cartilage cells, fat cells, and other kinds of connective tissue cells such as those in tendons. Neural stem cells in the brain give rise to its three major cell types: nerve cells (neurons) and two categories of nonneuronal cells, astrocytes and oligodendrocytes. Skin stem cells occur in the basal layer of the epidermis and at the base of hair follicles. The epidermal stem cells give rise to keratinocytes, which migrate to the surface of the skin and form a protective layer. The follicular stem cells can give rise to both the hair follicle and the epidermis.

Stem cells in adult mammalian tissues are rare and difficult to isolate. There is considerable debate concerning the plasticity of stem cells in adults. Plasticity is the ability of multipotent cells to exhibit pluripotency, such as the capacity of hematopoietic stem cells to differentiate into neurons.

Behavior in Cell Culture

During the 1980's, researchers first established in vitro culture conditions that allowed embryonic stem cells to divide without differentiating. Embryonic stem cells are relatively easy to grow in culture but appear to be genetically unstable; mice cloned from embryonic stem cells by nuclear transfer suffered many genetic defects as a result of the genetic instability of the embryonic stem cells. As embryonic stem cells divide in culture, they lose the tags that tell an imprinted gene to be either turned on or turned off during development. Researchers have found that even clones made from sister stem cells show differences in their gene expression. However, these genetic changes, while having defined roles in fetal development, may have little significance in therapeutic uses, because the genes involved do not serve a critical role in adult differentiated cells.

Unlike embryonic stem cells, adult stem cells do not divide prolifically in culture. When these stem cells do divide in culture, their division is unlike that of most cells. Generally, when a cell divides in culture, the two daughter cells produced are identical in appearance as well as in patterns of gene expression. However, when stem cells divide in culture, at least one of the daughter cells retains its stem cell culture while the other daughter cell is frequently a transit cell destined to produce a terminally differentiated lineage. The genes expressed in a stem cell and a transit cell are significantly different. Therefore a culture of adult stem cells may become heterogeneous in a short time.

Potential Therapeutic Issues

Although stem cells have significant use as models for early embryonic development, another major research thrust has been for therapeutic uses. Stem cell therapy has been limited almost exclusively to multipotent stem cells obtained from umbilical cord blood, bone marrow, or peripheral blood. These stem cells are most commonly used to assist in hematopoietic (blood) and immune system recovery following high-dose chemotherapy or radiation therapy for malignant and nonmalignant diseases such as leukemia and certain immune and genetic disorders. For stem cell transplants to succeed, the donated stem cells must repopulate or engraft the recipient's bone marrow, where they will provide a new source of essential blood and immune system cells.

In addition to the uses of stem cells in cancer treatment, the isolation and characterization of stem cells and in-depth study of their molecular and cellular biology may help scientists understand why cancer cells, which have certain properties of stem cells, survive despite very aggressive treatments. Once the cancer cell's ability to renew itself is understood, scientists can develop strategies for circumventing this property.

Research efforts are under way to improve and expand the use of stem cells in treating and potentially curing human diseases. Possible therapeutic uses of stem cells include treatment of autoimmune diseases such as muscular dystrophy, multiple sclerosis, and rheumatoid arthritis; repair of tissues damaged during stroke, spinal cord injury, or myocardial infarction; treatment of neurodegenerative diseases such as amyotrophic lateral sclerosis (ALS, commonly called Lou Gehrig's disease) and numerous neurological conditions such as Parkinson's, Huntington's, and Alzheimer's diseases; and replacement of insulin-secreting cells in diabetics.

Stem cells may also find use in the field of gene therapy, where a gene that provides a missing or necessary protein is introduced into an organ for a therapeutic effect. One of the most difficult problems in gene therapy studies has been the loss of expression (or insufficient expression) following introduction of the gene into more differentiated cells. Introduction of the gene into stem cells to achieve sufficient long-term expression would be a major advance. In addition, the stem cell is clearly a more versatile target cell for gene therapy, since it can be manipulated to become theoretically any tissue. A single gene transfer into a pluripotent stem cell could enable scientists to generate stem cells for blood, skin, liver, or even brain targets.

Ethical Issues

Stem cell research, particularly embryonic stem cell research, has unleashed a storm of controversy. One primary controversy surrounding the use of embryonic stem cells is based on the belief by opponents that a fertilized egg is fundamentally a human being with rights and interests that need to be protected. Those who oppose stem cell research do not want fetuses and fertilized eggs used for research purposes. Others accept the special status of an embryo as a potential human being yet argue that the respect due to the embryo increases as it develops and that this respect, in the early stages in particular, may properly be weighed against the potential benefits arising from the proposed research.

Another ethical issue concerns the method by which embryonic stem cells are obtained. Embryonic stem cells are isolated from two sources: surplus embryos produced by in vitro fertilization and embryos produced by somatic cell nuclear transfer (SCNT), often referred to as therapeutic cloning. In SCNT, genetic material from a cell in an adult's body is fused with an enucleated egg cell. With the right conditions, this new cell can then develop into an embryo from which stem cells could be harvested. Opponents argue that therapeutic cloning is the first step on the slippery slope to reproductive cloning, the use of SCNT to create a new adult organism. Proponents maintain that producing stem cells by SCNT using genetic material from the patient will eliminate the possibility of rejection when the resulting stem cells are returned to the patient.

IMPACT

Stem cell research, along with advances in genetics and cloning, gave impetus to a national policy on the use of stem cells. On August 9, 2001, President George W. Bush announced that federal funds could be used to support research using a limited number of human embryonic stem cell lines (about sixty) that had been derived before that date. However, there were no restrictions placed on the types of research that could be conducted on mouse embryonic stem cell lines and no federal law or policy prohibiting the private sector from isolating stem cells from human embryos. Several states introduced legislation to encourage research on stem cells taken from human embryos.

Neither reproductive cloning nor therapeutic cloning was forbidden by law in the United States. Congress has debated legislation in this area, however: One bill was proposed to ban both types of cloning, while an alternative proposal would ban only reproductive cloning. A number of states already have laws that ban human cloning for reproductive purposes, while a small number of states forbid cloning of embryos for stem cells as well.

The impact of policies to ban research must be weighed seriously in the global environment. On one hand, limits must be set and societal values must be debated in open public discussion. On the other, in a world in which national policies are imposed on scientific research, it is important to take into account the implications of different limitations—or lack of limitations—across the globe and how the dynamics of such different policies and laws may affect our society.

See also Chromosomes; Cloning; DNA Fingerprinting; DNA Sequencing; Double-Helix Model of DNA; Evolution; Gene-Chromosome Theory; Genetic Code; Human Evolution; Human Genome; Mendelian Genetics; Mitosis; Oncogenes; Population Genetics; Recombinant DNA Technology; Ribozymes; Viruses.

FURTHER READING

Holland, Suzanne, Karen Lebacqz, and Laurie Zoloth, eds. *The Human Embryonic Stem Cell Debate: Science, Ethics, and Public Policy (Basic Bioethics)*. Cambridge, Mass.: MIT Press, 2001.

Kaji, Eugene H., and Jeffrey M. Leiden. "Gene and Stem Cell Therapies." *Journal of the American Medical Association* 285, no. 5 (2001): 545-550.

Kiessling, Ann, and Scott C. Anderson. *Human Embryonic Stem Cells: An Introduction to the Science and Therapeutic Potential*. Boston: Jones and Bartlett, 2003.

Marshak, Daniel R., Richard L. Gardner, and David Gottlieb, eds. *Stem Cell Biology*. Woodbury, N.Y.: Cold Spring Harbor Laboratory Press, 2002.

Parson, Ann B. *The Proteus Effect: Stem Cells and Their Promise*. Washington, D.C.: National Academices Press, 2004.

Rao, Mahendra S., ed. *Stem Cells and CNS Development*. Totowa, N.J.: Humana Press, 2001.

—*Lisa M. Sardinia*

STONEHENGE

THE SCIENCE: English physician William Stukeley spent summers between 1719 and 1724 examining Stonehenge, Avebury, and related sites, producing exceptional notes but wrongly believing that Druids built the monuments. Nevertheless, his systematic method of investigation became a model for archaeological fieldwork.

THE SCIENTISTS:
William Stukeley (1687-1765), English antiquarian and archaeologist
John Aubrey (1626-1697), English antiquarian who discovered Avebury

EARLY THEORIES

The Wiltshire prehistoric sites of Avebury and Stonehenge are significant both in their construction and in the way they are situated in the landscape. Stonehenge is the ruin of a single building. An earthen embankment surrounded by a circular excavation ditch defines the site, although additional megaliths and earthworks lie outside the circle. In contrast with the compact area of Stonehenge, Avebury is a complex that covers several square miles, with a main circular bank and ditch and lined with megaliths, delimiting the original 30-acre site. In the eighteenth century, stones were dispersed among houses, gardens, and fields, making the layout difficult to discern. Avebury was not recognized as a human-made complex until 1649, when antiquarian John Aubrey discovered it during a hunting trip.

Although the site at Avebury had gone undetected, speculation about Stonehenge abounded for centuries. Noticed since medieval times, there were conflicting theories about its origin. In the early 1600's, poet and antiquarian Edmund Bolton credited its construction to the legendary first century military and rebel leader, Queen Boudicca. English architect Inigo Jones, who made the first known architectural study of the site, believed that it was a temple built by the Romans. Later in the seventeenth century, Walter Charleton, physician to King Charles II, claimed it was built by Danes. Aubrey, after his discovery of Avebury, investigated both monuments and believed that they were of Druid origin.

STUKELEY'S FIELDWORK

William Stukeley first visited Stonehenge and Avebury (which he called, collectively, Abury) in 1719. Although he was a trained physician, he pursued studies in theology, science, and antiquities. A member of the Society of Antiquaries and a fellow of the Royal Society, he was a colleague of the most gifted individuals of eighteenth century England. He explored the English countryside observing and recording ancient monuments.

Stukeley was familiar with Aubrey's then unpublished *Monumenta Britannica: Or, A Miscellany of British Antiquities* (1980-1982), which recorded Aubrey's theories along with his observations and measurements of Avebury and Stonehenge. Like Aubrey, Stukeley believed that the monuments were built in pre-Roman times. Furthermore, he felt that his theory could be proven. He speculated that compilation of data about the circles and other ancient sites could provide information not obtainable from written sources.

There are few particulars about Stukeley's visits to Avebury and Stonehenge in 1719 and 1720, but from 1721 to 1724, after he decided to develop a typology of ancient monuments, he detailed his studies. Although Aubrey's work provided an underacknowledged precedent, it was not as encompassing as the project undertaken by Stukeley. Each summer he conducted fieldwork, living on site. His techniques of observation, accurate measurement, and detailed recording accompanied by carefully executed drawings have led to Stukeley being recognized as the foremost figure in eighteenth century English archaeology.

Close observation was a key element in developing his typological study, as was evident in his *Itinerarium curiosum: Or, An Account of the Antiquitys and Remarkable Curiositys in Nature or Art* (1724). Here he noted common building characteristics, such as placement of upright stones in a circular pattern on elevated ground with a surrounding ditch, a surrounding plain,

Stonehenge. (Library of Congress)

and an avenue of approach. Through this typology he wanted to show that Stonehenge and Avebury had the same provenance as other stone temples in England.

Much of his work was without precedent. He developed a vocabulary to describe his findings; he coined the term "trilithon," for example, to describe two upright stones supporting a lintel. He pioneered the field of astroarchaeological studies by being the first to note that Stonehenge was astronomically aligned: The site's assumed entrance marks the point of sunrise on the summer solstice. In 1721 he was the first to discern a raised area, which he called the "avenue," extending from the entrance to Stonehenge toward the River Avon; although the lining stones were gone, he measured placement intervals after observing sockets remaining in the uncultivated ground. Also, in 1723, he discovered at Stonehenge a shallow enclosure of parallel ditches measuring 2 miles in length; he called this the "cursus," speculating that it was an ancient racetrack. At Avebury he discovered similar stone-lined constructions leading toward West Kennet and Beckhampton.

To establish his typology, Stukeley needed measurements from many ancient sites. He stressed precision, believing valid conclusions could be drawn only from accurate comparisons. In 1723 he and Lord Winchelsea took two thousand measurements at Stonehenge, attempting to detect a common, indigenous standard of measurement, which Stukeley called "Druid's cubit," to prove pre-Roman origins of megalithic sites. Through reading and correspondence he also compiled measurements of stone circles located outside the sphere of Roman occupation.

In addition to recording his observations and measurements, Stukeley developed excavation techniques, which he compared to anatomical dissection. In 1722 and 1723 he and Lord Pembroke excavated Bronze Age barrows around Stonehenge. Stukeley's careful technique surpassed anything undertaken prior to that time. He noted that stratigraphy had the potential to establish chronology. He studied construction of barrows and their funerary contents, made precise notes, and carefully drew a cross-section diagram, which was the first such visual record in British archaeology.

Drawings and diagrams played an important role in his fieldwork. From 1721 to 1723 he diagramed the main circles within the great ditch at Avebury and also indicated the avenue of standing stones leading toward West Kennet. The avenue terminated in a double circle of standing stones called the "sanctuary" by local villagers. Stukeley then recorded what remained of the sanctuary and marked discernible sites of destroyed stones.

It has been suggested that the ongoing destruction at Avebury and Stonehenge induced Stukeley to prepare records before the monuments were lost. In the Middle Ages, megaliths often were regarded as pagan relics and were buried. In the eighteenth century, the Avebury site was used as a quarry for building stone. Stukeley also noted that visitors hammered off pieces of the monuments for souvenirs. The owner of Avebury Manor destroyed part of the site's embankment to build a barn. Each year, cultivation further eliminated features of the prehistoric landscape.

After he was ordained into the Church of England in 1729, Stukeley became increasingly conjectural in interpreting the past. Responding to the perceived threat of Enlightenment secularism, he romanticized Druids and postulated that the Church of England was prefigured in their ancient religion. In three works—*Palaeographia sacra: Or, Discourses on Monuments of Antiquity that Relate to Sacred History* (1736); *Stonehenge: A Temple Restor'd to the British Druids* (1740); and *Abury: A Temple of the British Druids* (1743)—he mixed religious speculation with his scientific fieldwork.

IMPACT

William Stukeley's writing reflected the dual nature of thought in the eighteenth century, which incorporated rational-scientific as well as religious-romantic ideas. His linking of Avebury and Stonehenge with Druidism became an enduring fallacy that was expressed in the Romantic tradition in English literature of the late eighteenth and early nineteenth centuries. Poetry by Thomas Gray and William Collins reflected a Druidical revival, as did works by the artist and poet William Blake.

Because Stukeley's scientific studies were intermingled with his Druidic theories, the accuracy of his field surveys has been questioned. Subsequent studies at Stonehenge and Avebury, however, have validated his work, which has provided a valuable record of historic sites before they were subjected to additional ravages of agricultural and economic development. Early aerial photography of the 1920's corroborated Stukeley's observations of the avenue at Stonehenge. Excavations in 1930 confirmed the existence of Avebury's sanctuary, which was destroyed shortly after Stukeley's documentation. The frontispiece to *Abury* provided accurate site information that was used in Alexander Keiller's excavations of 1934 to 1938. The Beckhampton Avenue stones at Avebury no longer exist, but recent excavations substantiate Stukeley's findings.

Stukeley was a key figure in bridging antiquarianism and the emerging science of archaeology. His pioneering work, although based on Aubrey's early techniques, provided the most thorough, systematic studies of Avebury and Stonehenge attempted before the nineteenth century. His careful observations, measurements, and diagramed descriptions were significant components in the development of the field of archaeology. He compiled enough data to recognize these structures as representing a larger group of monuments scattered across Britain. He correctly conceived of these sites as prehistoric sanctuaries. At a time when scholars still used Old Testament chronologies for establishing historical dates, he set about proving that native Britons created the monuments in pre-Roman times. He was among the first to recognize their historic value and to express concern over their preservation. Contemporary analysis of Stukeley's detailed records reveals how much has been lost from the sites in the past two centuries, either taken or destroyed, or both. Historians and archaeologists are indebted to Stukeley for charting the course and following the traces.

See also Dead Sea Scrolls; Pompeii; Rosetta Stone; Troy.

Further Reading

Chippindale, Christopher. *Stonehenge Complete*. New York: Thames and Hudson, 2004.

Haycock, David. *William Stukeley: Science, Religion, and Archaeology in Eighteenth-Century England*. Woodbridge, England: Boydell Press, 2002.

Piggott, Stuart. *William Stukeley: An Eighteenth-Century Antiquary*. New York: Thames and Hudson, 1985.

Stukeley, William. *The Commentarys, Diary, and Common-Place Book and Selected Letters of William Stukeley*. London: Doppler Press, 1980.

_____. *"Stonehenge, a Temple Restor'd to the British Druids"* and *"Abury, a Temple of the British Druids."* New York: Garland, 1984.

—*Cassandra Lee Tellier*

STRATOSPHERE AND TROPOSPHERE

THE SCIENCE: Based on experimental balloon measurements of atmospheric temperature versus height, Léon Teisserenc de Bort discovered the stratosphere's and troposphere's vertical layering on the basis of thermal inversion.

THE SCIENTISTS:

Léon Teisserenc de Bort (1855-1913), French physicist and meteorologist
Richard Assmann (1845-1918), German physicist and meteorologist

TEMPERATURE AND ALTITUDE

The details of the rate of change of atmospheric temperature versus height have been of basic importance for many years in trying to determine and predict the processes governing weather. For example, the variation of wind with height also depends upon vertical temperature variation.

Until 1883, the body of air above the Earth's surface was considered generally a uniform body. Then, violent eruption of the volcano Krakatoa in the Java Sea in 1883 produced abnormally high atmospheric concentrations of dust, implying the existence of higher-level global temperature and wind patterns. William Morris Davis's *Elementary Meteorology* (1894) is representative of knowledge of the upper atmosphere before large-scale kite and balloon sondings. Davis simply divided the Earth into geosphere (rock), hydrosphere (water), and atmosphere (air). An empirical formula for atmospheric temperature gradient was developed by Austrian meteorologist Julius Ferdinand von Hann in 1874, based on indirect atmospheric measures such as astronomical observations of the duration of twilight and of meteor burns. Davis proposed that successive isobaric (equipressure) surfaces were separated by greater and greater distances indefinitely, out into space. The general distribution of temperature with elevation was simply illustrated as a nearly linear decreasing function.

BALLOON ASCENTS

Manned balloon ascents to measure upper air temperature were first undertaken by John Jeffries and François Blanchard in 1784 and subsequently by Jean-Baptiste Biot and Joseph-Louis Gay-Lussac in 1804, and continued in England in 1852. Factors influencing balloon performance included the excess of buoyancy forces over balloon gross weight (including human observers) and the maximum size to which the balloon's silk or India rubber envelope would expand in response to decreasing atmospheric pressure. These factors control both maximum ascent ceiling and ascent rate. The need for light gases, such as hydrogen or helium, is to keep the balloon's envelope sufficiently distended. The buoyancy force, which arises from Archimedes' principle, is equal to the air mass displaced by the balloon. As the balloon rises, the air density falls by a factor of about ten for every 10 kilometers of ascent; therefore, the balloon's envelope expands in exact proportion to falling density.

Prior to 1890, balloon observations were, for the most part, limited to heights of only a few kilometers by human oxygen consumption, recording mainly local rather than regional or global temperature behaviors. The first attempts at global isothermal charts were published by Hann in Vienna and Alexander Buchan in Edinburgh in 1887 and 1889, respectively. To overcome the human limitation, kites were first employed by Cleveland Abbe in studying winds under a thunder cloud at the Blue Hill Observatory in Massachusetts. Nevertheless, for technical reasons, the maximum heights attained by kites were only about 8 kilometers.

Because of proven dangers to human life in high ascents, small free rubber balloons carrying recently developed self-recording temperature and pressure recorders were first deployed in 1893 by French aeronomist Georges Besançon and were rapidly adopted elsewhere for meteorological observations. When atmospheric visibility is sufficiently good, larger meteorological balloons could be followed visually by theodolites to obtain supplementary wind direction data. Theodolites are grid-mounted survey telescopes permitting measurement of height and angular motion. These various observations demonstrated that to at least about 9,000 meters, temperature decreased in a fairly uniform fashion at a rate of about 1 degree Celsius per every 180 meters risen.

Teisserenc de Bort's Sondings

After extensive work in Europe and North Africa with the French government undertaking barometric and other weather observations, in 1897, Léon Teisserenc de Bort founded his own private aeronomic observatory at Trappes near Paris. Earlier, Teisserenc de Bort had pioneered self-recording temperature and barometric pressure sensors; the physicist Richard Assmann developed the first self-recording hygrometer to measure atmospheric humidity. Using hydrogen-filled balloons specially designed for rapid and near vertical ascents, Teisserenc de Bort named his surveys "soundings" or "sondings," in analogy to bathymetric depth soundings by sonde-line or acoustic sound at sea. A critical factor was sufficient protection of thermometers from direct solar radiation, as well as recorders that could respond to changing temperature faster than the balloon would rise.

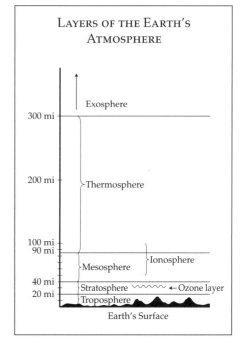

LAYERS OF THE EARTH'S ATMOSPHERE

In April, 1898, Teisserenc de Bort, using his improved apparatus, began a long series of regular balloon sondings from Trappes, France. Among other details, he soon discovered unusual temperature records, first believed to be instrument errors, of constant or even increasing temperature conditions from the extreme upper limits of his balloon's ascents. After precluding instrument error and repeating many measurements, in 1899, he published a report indicating that temperatures at heights above 0.1 atmospheric pressure (100 millibars) cease to decline with altitude but remain constant over a specific height interval, thereafter slowly increasing.

In his papers of 1904 in the noted French journal *Comptes rendus physique* and his own *Travaux scientifiques de l'observatoire de météorologie de Trappes*, Teisserenc de Bort gave mean temperatures versus height measured at Trappes between between 1899 and 1903. Out of 581 balloon ascents, 141 attained temperature "isothermal" and "inverted" measurements at height records of 14 kilometers or more. His data showed that there is a slow temperature decrease up to about 2 kilometers above sea level. This is followed by a more rapid decrease up to about 10 kilometers. A very slow or total lack of decrease was measured between 11 and 14 kilometers (with an ambient temperature of about –55° Celsius). He called this the "thermal" zone or boundary.

Teisserenc de Bort's observations were almost concurrently confirmed by Assmann's independent series of ascents from Berlin. Assmann and Artur Berson, beginning in 1887, undertook a more extensive series of upper atmospheric soundings, under the aegis of the Prussian Meteorological Office and Aeronautical Section of the German Army, and later as an independent scientific station at Lindenberg. The details of their seventy ascents between 1887 to 1889 were the first published aeronometric measurements of temperature for several locations, in 1900, and thereafter published regularly in the German journal *Das Wetter*. From a particularly long series of kite soundings from Berlin between October, 1902, and December, 1903, Assmann showed that atmospheric temperature is much more variable at 6 or 7 kilometers height than at ground surface. The effects of diurnal and seasonal changes on upper-level temperatures were also measured. Following the systematic planned simultaneous ascents from many European cities between 1895 and 1899, Assmann assembled a data base of more than one thousand of his own observations, with 581 of Teisserenc de Bort, and others from England, Holland, and the Soviet Union, enabling him to compute monthly and annual temperature and wind velocity averages of many altitudes between 0 and 11 kilometers over central Europe. Assmann also argued that at about 12 kilometers, the upper limit for cirrus clouds, temperature remains constant and later increases slowly. The atmospheric region above these heights of constant temperature was called the stratosphere, the lower region nearest the ground was called the troposphere, and the transition zone was called the tropopause. The mesosphere and thermosphere are above the stratosphere.

IMPACT

Meteorologic sonding heights of more than 25 kilometers were achieved in France and Belgium between 1905 and 1907. The Fifth Conference of the International Committee on Scientific Aeronautics at Milan in 1906 saw an increasing number of measurements confirming the temperature results of Teisserenc de Bort and Assmann, notably kite ascents from 1904 to 1905 from the Soviet Union. These data established that above a height that geographically varied from about 18 kilometers near the equator to about 11 kilometers at 50° north latitude to only about 6 kilometers at the poles, atmospheric temperature remained approximately constant over a certain level. (The English meteorologist W. Dines subsequently showed that the stratosphere is high and cold over high pressure and low and warm over low pressure.)

As soon as diverse independent observations had established the troposphere/ tropopause/stratosphere, many efforts were made to explain rigorously the occurrence of stationary upper-level discontinuities on the basis of the rapidly developing hydrothermodynamics of Vilhelm Bjerknes, Ludwig Prandtl, and others—initially, however, with only very limited success. Finally, W. Humphreys in the United States (*Astrophysical Journal*, vol. 29, 1909) and F. Gold in England (*Proceedings of the Royal Society*, vol. 82, 1909) published what became essentially the generally accepted explanation. In both approaches, it was recognized that it is necessary to consider the thermodynamic balance between absorbed and reemitted solar radiation. Humphreys' account is less mathematical but equivalent to Gold's. Briefly, since the average annual temperature in the atmosphere at any location had been shown experimentally not to vary greatly, Humphreys concluded that the absorption of solar radiation is equal basically to the net outgoing reradiation by Earth (discovered previously by S. Langley), using a simple thermodynamic "blackbody" model. Humphreys concluded that the isothermal/tropopause zone marks the limit of vertical thermal convection and, from this, correctly deduced that the above-lying layers are warmed almost entirely by direct solar radiation (later shown to be dependent upon atmospheric ozone). The increasing temperature trend was shown later to be caused directly by the heat released during the interaction between incoming ultraviolet radiation and atmospheric ozone molecules.

Further direct and indirect studies of the stratosphere and troposphere continued by a variety of means. In studies of ground versus air waves from earthquakes by Emil Wiechert in 1904, and later during World War I, it was noted that loud noises could be heard occasionally at distances ranging from 150 kilometers to more than 400 kilometers from their source, even when observers near the source could barely hear the sounds. Between 1928 and 1931, H. Benndorf, P. Duckert, and O. Meissner made recordings of seismic-acoustic wave propagation in atmospheric temperature inversions associated with the troposphere. Sound waves are bent gradually or refracted resulting from the increased velocity of sound in air resulting from a gradient of rising temperature at about 35 to 60 kilometers height. These observations provided another method of estimating temperature then inaccessible to aircraft, balloon, and kite soundings. In 1926, G. Dobson and F. Lindemann employed data from hundreds of meteor burn observations to extrapolation temperature, pressure, and chemical observations to heights of up to 160 kilometers, confirmed by V-2 flights during and after World War II.

Subsequent studies of the stratosphere by Earth-orbiting satellites include the mapping of the (polar) jet streams and the twenty-six-month quasi-biennial cycle. The original motivation and basis for these and other studies, however, remain the methods and results of Teisserenc de Bort and Assmann.

See also Atmospheric Circulation; Atmospheric Pressure; Chlorofluorocarbons; Global Warming; Ionosphere; Ozone Hole; Van Allen Radiation Belts; Weather Fronts.

FURTHER READING

Anthes, Richard A., et al. *The Atmosphere*. Columbus, Ohio: Charles Merrill, 1975.
Davis, William Morris. *Elementary Meteorology*. Boston: Ginn, 1894.
Goody, Richard M. *The Physics of the Stratosphere*. London: Cambridge University Press, 1954.
Humphreys, W. J. *The Physics of the Air*. New York: Dover Books, 1964.
_____. "Vertical Temperature Gradient of the Atmosphere, Especially in the Region of the Upper Inversion." *Astrophysical Journal* 29 (1909): 14-26.
Massey, Harrie Stewart Wilson. *The Middle Atmosphere as Observed from Balloons, Rockets, and Satellites*. London: Royal Society, 1980.

—*Gerardo G. Tango*

STREPTOMYCIN

THE SCIENCE: Selman Abraham Waksman searched for an antibacterial substance in soil microorganisms, discovering eighteen antibiotics, including streptomycin, the first drug effective against tuberculosis.

THE SCIENTISTS:

Selman Abraham Waksman (1888-1973), Soviet-born American soil microbiologist and winner of the 1952 Nobel Prize in Physiology or Medicine
René Dubos (1901-1982), French-born American microbiologist
William Hugh Feldman (1892-1974), American pathologist

CURING COWS

Some microbiologists in the late nineteenth century believed that a struggle for survival occurred in the microbial world. They thought that microbes might contain substances that inhibited the growth of other microbes. There were attempts to isolate chemotherapeutic agents from such microbial substances as molds and bacteria, but the field was abandoned in the early twentieth century until the reawakening of interest in such agents by René Dubos in the 1930's.

Dubos was a student of Selman Abraham Waksman, whose area of expertise was the population of microorganisms that inhabit the soil. Waksman specialized in one type of soil microbe, the actinomycetes, organisms intermediate between bacteria and fungi. His research included a study on how the tubercle bacillus fared when introduced into soil. From 1932 to 1935, Waksman established that the germ could not survive because of the antagonism of soil microbes. The finding substantiated the fact that pathogenic germs do not survive when introduced into soil. At the time, his finding did not seem to lead to anything new; it was only another example of microbes inhibiting other microbes.

Dubos wondered what would happen if soil were enriched with pathogenic germs. He pondered if perhaps their presence would encourage soil microbes antagonistic to them to flourish. In February, 1939, Dubos announced that he had tracked down such an antagonistic microorganism, *Bacillus brevis*, and from it had isolated two antibacterial substances, tyrocidine and gramicidin. The latter proved to be the first true antibiotic drug, attacking pneumococcus, staphylococcus, and streptococcus germs. Too toxic for human therapy, it became useful in treating animals. It aroused public interest when, at the 1939 New York World's Fair, sixteen of the Borden cow herd developed a streptococcal udder infection and gramicidin cured twelve of the cows of the bacteria.

CURING TUBERCULOSIS

Dubos's discovery alerted scientists to the possibility of finding other powerful drugs in microorganisms, and the central figure in exploiting this field was Waksman. He seized on Dubos's work and converted his research on soil actinomycetes into a search for the antibacterial substances in them. The actinomycetes proved to be the most fertile source for antibiotics. Waksman coined and defined the term "antibiotic" in 1941 to describe the novel drugs found in microbes. He developed soil enrichment methods and discovered eighteen antibiotics between 1940 and 1958. He cultured thousands of soil microbes in artificial media and screened them for activity. The promising ones were then chemically processed to isolate the antibiotics.

Streptomycin was the most important of Waksman's discoveries. In September, 1943, with his students Elizabeth Bugie and Albert Schatz, he isolated a soil actinomycete, *Streptomyces griseus*, which contained an antibiotic he named "streptomycin." It was antagonistic to certain types of bacteria. His report appeared in January, 1944, and two months later, another article claimed that streptomycin was active against the deadly tubercle germ, *Mycobacterium tuberculosis*.

In the 1940's, tuberculosis was not fully under control. There was no cure, only prolonged bed rest and a regimen of nutritious food. The tubercle germ could invade any organ of the body, and in its various forms, the disease took a horrifying toll. A diagnosis of tuberculosis entailed lifelong invalidism, and patients died because the available treatment was so limited.

Selman Abraham Waksman. (The Nobel Foundation)

As the search for a cure progressed, the medical world took notice of the clinical tests conducted by William Hugh Feldman and H. Corwin Hinshaw at the Mayo Clinic. They had been investigating the chemotherapy of tuberculosis since the 1930's. Many scientists believed that such therapy was unattainable, but Feldman and Hinshaw refused to accept this verdict. They worked with sulfa drugs and sulfones and found some effect in suppressing the growth of tubercle bacilli, but not their eradication. Feldman had visited Waksman before the discovery of streptomycin and indicated a desire to try any promising antibiotics.

When Waksman found antitubercular effects in 1944, he wrote at once to Feldman to offer streptomycin for his studies. Feldman and Hinshaw had developed a practical system to determine the ability of a drug to slow the course of tuberculosis in guinea pigs. They used streptomycin on guinea pigs inoculated with the tubercle germ. In December, 1944, they issued their first report. The tests revealed streptomycin's ability to reverse the lethal course of the inoculations, and they concluded that it was highly effective in inhibiting the germ, exerting a striking suppressive effect, and was well tolerated by the animals.

Feldman and Hinshaw were now ready to test human patients. Hinshaw enlisted two physicians from a nearby sanatorium. On November 20, 1944, and for the next six months, a twenty-one-year-old woman with advanced pulmonary tuberculosis received streptomycin. In June of 1945, she was discharged, her tuberculosis arrested; she married eventually and reared three children.

This happy ending was followed by many more. Feldman and Hinshaw deserve the credit for proving that streptomycin could be used against tuberculosis. They demonstrated its value in carefully conducted trials. Some observers believe that they should have shared the 1952 Nobel Prize with Waksman.

IMPACT

Waksman did more than discover a major antibiotic. His work encouraged others to attempt to isolate new antibiotics by means of screening programs similar to those he devised. The 1950's witnessed a large increase in the number of antibiotics, and antibiotics became a large industry with total production of more than nine million pounds in 1955.

Streptomycin was not perfect. As early as 1946, reports appeared on the resistance of bacilli to the drug. Such resistant strains could be responsible for the failure of therapy. New drugs came to the rescue; Swedish investigators found that a drug consisting of para-aminosalicylic acid would inhibit the tubercle bacillus, although not as effectively as streptomycin. In 1949, the Veterans Administration combined the two drugs. After that, "combination" therapy proved to be the key to the future of chemotherapy, as the combination delayed the appearance of resistant strains. By 1970, using available drugs, and by the judicious use of combinations, physicians could achieve recovery in nearly all cases of pulmonary tuberculosis.

See also Anesthesia; Antisepsis; Aspirin; Contagion; Diphtheria Vaccine; Germ Theory; Hybridomas; Immunology; Penicillin; Polio Vaccine: Sabin; Polio Vaccine: Salk; Schick Test; Smallpox Vaccination; Viruses; Yellow Fever Vaccine.

FURTHER READING

Dowling, Harry F. *Fighting Infection: Conquests of the Twentieth Century.* Cambridge, Mass.: Harvard University Press, 1977.
Epstein, Samuel, and Beryl Williams. *Miracles from Microbes: The Road to Streptomycin.* New Brunswick, N.J.: Rutgers University Press, 1946.
Lappé, Marc. *Germs That Won't Die.* Garden City, N.Y.: Doubleday, 1982.
Lechevalier, Hubert A., and Morris Solotorovsky. *Three Centuries of Microbiology.* New York: McGraw-Hill, 1965.
Waksman, Selman A. *The Conquest of Tuberculosis.* Berkeley: University of California Press, 1964.
_____. *My Life with the Microbes.* New Brunswick, N.J.: Rutgers University Press, 1954.

—Albert B. Costa

STRING THEORY

THE SCIENCE: The theory of cosmic strings provided a workable explanation of how matter formed into stars, galaxies, and clusters.

THE SCIENTISTS:
Tom Kibble (b. 1932), English physicist
Neil Turok, English physicist
Andreas Albrecht (b. 1927), American quantum chemist
Edward Witten (b. 1951), American physicist

THE MOMENT AFTER CREATION

According to the "big bang" theory, the universe began with a colossal explosion fifteen to twenty billion years ago. Modern advances in particle physics and other branches of the physical sciences have allowed scientists to formulate theories that describe the events immediately following the big bang. According to one theory, from the moment of creation until a point some 10^{-43} second later, all four of the forces of nature consisted of one "superforce." The universe consisted of energy; there were no elementary particles. Physicists refer to this state as one of symmetry. In other words, the universe at that time would have appeared to have the same properties in all directions. As minute increments of time passed, the symmetry was broken as individual forces began to appear. First came gravity, and then the strong nuclear force. At about 10^{-12} second, the weak force and electromagnetism began to exist as independent forces. The appearance of these forces made possible first the formation of elementary particles and then, within minutes, the first atomic nuclei. After much expansion and cooling of the universe, the first atoms were formed. Physicists estimate that this latter event took place about 700,000 years after the initial explosion in which the universe was created.

Prior to the formation of the first atoms, the vast number of free electrons in the universe interacted with light emitted at the instant of the big bang. After most of the electrons had become involved in the formation of atoms, matter and light were decoupled and the universe became transparent to radiation. At this juncture, reduced light pressure allowed bits of matter to form larger masses.

LIKE CRACKS IN ICE

According to the first theories of galactic formation, gravitational forces acting in the early universe caused matter to form lumps. These lumps, in turn, attracted great clouds of dust, and from these huge rotating masses, individual stars were born. Stars that were formed close to one another remained gravitationally bound and formed huge multibillion-star assemblages called galaxies. Individual galaxies were attracted by gravity to form clusters, and clusters were bound to superclusters.

The problem with this theory is that it does not explain why or how matter formed into lumps in the first place. Since cosmic radiation—the remnant of the big bang fireball—is the same in intensity from all parts of the sky, it is difficult to accept the idea that there may have been irregularities in the explosion that could have caused some unevenness in the distribution of matter.

In 1976, physicist Tom Kibble, working at the Imperial College in London, was considering the possible effects of modern theories of unified fields on the universe. He was particularly concerned with that fraction of a second after the big bang when the forces (fields) began to assume their separate identities. His mathematical model suggested that shortly after the big bang, the rapidly cooling universe developed flaws that appeared to be stringlike in nature. This rapid cooling of the universe would produce what is called a "phase transition," which is analogous to the cracks and other flaws that are formed when water is frozen into ice. Kibble's strings were described as slender strands of highly concentrated mass-energy. These remnants of the original fireball, according to the theory, are much thinner in diameter than a proton and as long as the known universe. A segment of a cosmic string 1.6 kilometers long would weigh more than the entire Earth. This large mass suggests that strings must have been formed early in the history of the universe, when there was an excess of energy.

Tom Kibble. (Imperial College of Science, Technology, and Medicine)

IMPACT

Computer simulations conducted by Neil Turok and Andreas Albrecht indicate that as the universe expanded and rapidly cooled immediately after the big bang, defects in space-time formed long, continuous chains. Within these chains or strings, symmetry still exists. The forces of nature exist as one force, and as a result, there are no atomic particles. As the universe expands, the strings evolve. Rapid vibrations within any one string may cause portions of that string to overlap. When this occurs, the loop that has been formed breaks away from the string. These loops may be of any size, from microscopic to several light-years across.

According to the theory of cosmic strings, the loops undergo rapid oscillations. These oscillations, the speed of which may approach the speed of light, cause the emission of gravitational waves. These waves, which were predicted by Albert Einstein's general theory of relativity, are ripples in the fabric of space-time. As a string radiates this energy, it eventually shrinks and disappears. It has been estimated that a loop of cosmic string 1,000 light-years in circumference would radiate away in 10 to 100 million years.

Scientists wonder whether any strings remain in the universe. Researchers working on string theory have determined that the smallest loop that could have been formed in the primeval universe and could still exist must have had an initial diameter of at least one million light-years. It is also theorized that any currently existing strings would be widely dispersed, perhaps as much as one billion light-years from Earth.

A modification of cosmic string theory by Edward Witten suggests that strings might be superconductors of electricity. It has been calculated that currents as great as 100 quintillion amperes could be induced. The flow of electrical current produces a magnetic field, so strings should be surrounded by intense fields. Particles trapped and accelerated within these fields would glow. The observation of radiation from these particles might one day provide the first evidence of the existence of cosmic strings.

See also Big Bang; Black Holes; Brahe's Supernova; Cassini-Huygens Mission; Cepheid Variables; Chandrasekhar Limit; Copernican Revolution; Cosmic Microwave Background Radiation; Cosmic Rays; Expanding Universe; Extrasolar Planets; Galactic Superclusters; Galaxies; Galileo Mission; Gamma-Ray Bursts; Gravitation: Einstein; Greek Astronomy; Halley's Comet; Heliocentric Universe; Herschel's Telescope; Hubble Space Telescope; Inflationary Model of the Universe; Jupiter's Great Red Spot; Kepler's Laws of Planetary Motion; Mars Exploration Rovers; Mass Extinctions; Mayan Astronomy; Moon Landing; Nebular Hypothesis; Neutron Stars; Oort Cloud; Planetary Formation; Pluto; Pulsars; Quasars; Radio Astronomy; Radio Galaxies; Radio Maps of the Universe; Saturn's Rings; Solar Wind; Speed of Light; Stellar Evolution; Van Allen Radiation Belts; Very Long Baseline Interferometry; Voyager Missions; Wilkinson Microwave Anisotropy Probe; X-Ray Astronomy.

FURTHER READING

Abell, George O., David Morrison, and Sidney C. Wolff. *Realm of the Universe*. New York: Saunders College Publishing, 1987.
Hartman, William K. *The Cosmic Voyage: Through Time and Space*. Belmont, Calif.: Wadsworth, 1990.
Press, William H., and David N. Spergel. "Cosmic Strings: Topological Fossils of the Hot Big Bang." *Physics Today* 42 (March, 1989): 29-35.
Trefil, James S. *The Dark Side of the Universe. Searching for the Outer Limits of the Cosmos*. New York: Charles Scribner's Sons, 1988.
Vilenkin, Alexander. "Cosmic Strings." *Scientific American* 257 (December, 1987): 94-98.

—David W. Maguire

SUPERCONDUCTIVITY

THE SCIENCE: John Bardeen, Leon N. Cooper, and John Robert Schrieffer were the first physicists to explain how some metals, as they approach absolute zero (–237.59° Celsius), lose their resistance to electricity.

THE SCIENTISTS:

John Bardeen (1908-1991), American physicist
Leon N. Cooper (b. 1930), American physicist
Fritz Wolfgang London (1900-1954), American physicist
Heinz London (1907-1970), American physicist
Heike Kamerlingh Onnes (1853-1926), Dutch physicist
John Robert Schrieffer (b. 1931), graduate student in physics

A SCIENTIFIC MIRACLE

When an electric current is run through a piece of metal, a considerable amount of the energy is lost to what is called "electrical resistance." Different metals have different resistances. In 1911, Heike Kamerlingh Onnes, a Dutch physicist who later won a Nobel Prize in Physics, made a startling discovery. He found that when the temperature of mercury was lowered almost to absolute zero (–273.15° Celsius), it seemed to lose all of its electrical resistance. Kamerlingh Onnes had discovered superconductivity. For the next fifty years, scientists would repeat this experiment with other metals with the same results, but no one was able to explain it.

John Bardeen first became interested in superconductivity in 1938, when he read David Shoenberg's new book *Super-Conductivity*. He had already heard of the report given by Fritz London and Heinz London to a meeting of the Royal Society that established a link between superconductivity and quantum mechanics, something else that physicists were only beginning to understand. By 1940, Bardeen had begun to formulate his own explanation for superconductivity. Unfortunately, his thinking was sidetracked by World War II. Between 1941 and 1945, Bardeen worked at the Naval Ordnance Laboratory, doing research on the transistor. He was awarded his first Nobel Prize for this research in 1956.

Bardeen resumed his study of superconductivity in 1950, after some new discoveries helped to explain how electricity works. These discoveries turned out to be some of the missing pieces in the puzzle of superconductivity, though the puzzle still seemed almost impossible to solve. Bardeen decided that the only way to solve the puzzle was to break it up into smaller, less difficult pieces. In 1951, having become a professor of physics and engineering at the University of Illinois at Urbana-Champaign, he asked Leon N. Cooper, who had a doctorate in physics from Columbia University, to help him with some of the problems. In 1956, they were joined by John Robert Schrieffer, one of Bardeen's graduate students.

It was already known that twenty-six metals and ten alloys (combinations of metals) are superconductors at different temperatures. The highest temperature at which any of them became superconductors was −214.44° Celsius. As yet, there was no real evidence that superconductivity had to take place at such extremely low temperatures. Although using liquid nitrogen is a fairly cheap way to bring temperatures down close to absolute zero, Bardeen and his coworkers knew that superconductivity would never be truly practical unless it could be achieved at higher temperatures.

Cooper Pairs

As early as 1950, Bardeen had understood that the key to understanding superconductivity lay in the way in which electrons move and interact with one another. Earlier theories had supposed that superconductivity involved atomic vibration. Bardeen suggested possible ways of measuring the change in electrical resistance when superconducting temperatures had been reached. Until this time, there had been no instruments sensitive enough to measure this resistance.

By 1956, Cooper, Bardeen's assistant, had finally taken the first step toward solving the puzzle. He discovered that free electrons (electrons that are not bound to any one molecule) are attracted to each other in pairs during superconductivity. These pairs are often called "Cooper pairs." After a year of working with this discovery, Bardeen and his coworkers finally came up with a successful model for understanding superconductivity. They published this model in February, 1957, and followed it with supporting evidence for the next six months.

One way to view their model is to think of the electrons as people in a crowded railroad station. The people are squeezed together so tightly that they bump into anything that gets in their way—mostly other people. This is how electrons move when there is little or no current running through a metal.

When a current is introduced, it is as though the people on one side of the station were suddenly pushed very hard. The people in their immediate vicinity are pushed very hard as well, but the people who are farther away feel the push only slightly, and the people on the other side of the station may not feel anything. If electrons are weakly paired, however, as the Cooper pairs are, the resistance to any one pair from all the electrons that are not a part of that pair goes down about one hundred times. It is this state that comes close to superconductivity. Bardeen and his coworkers also discovered that besides lacking resistance, superconductors can also prevent magnetic fields from entering them.

IMPACT

For centuries, people have been seeking a way to use perpetual motion to create energy without using fuel. When Bardeen proposed his theoretical explanation of superconductivity, scientists began to think that the search was almost over. When the theory was announced, scientists immediately started to devise ways to put superconductivity to use. They thought of electrical power lines made of superconductive wire. Without any electrical resistance to stop much of the electricity from reaching the end of the line, electrical power plants would be much cheaper and more efficient. Scientists also suggested superfast trains that would be built on top of powerful magnets and hover over superconductive "rails." They imagined computers running hundreds of times faster than had been possible before the development of superconductors. These ideas and many more were made possible by the theory set forth by Bardeen, Cooper, and Schrieffer. Many ideas have already been put to work, and others, still unthought of, will be developed in the years to come.

See also Buckminsterfullerene; Electron Tunneling; Kelvin Temperature Scale; Liquid Helium; Quantized Hall Effect; Superconductivity at High Temperatures; Thermodynamics: Third Law.

Further Reading

Bogoliubov, Nikolai N., V. V. Tolmachev, and D. V. Shirkov. *A New Method in the Theory of Superconductivity*. New York: Consultants Bureau, 1959.
Hoddeson, Lillian. "John Bardeen and the Discoveries of the Transistor and the BCS Theory of Superconductivity." In *A Collection of John Bardeen's Publications on Semiconductors and Superconductivity*. Urbana: University of Illinois Press, 1988.
London, Fritz. *Superfluids*. New York: John Wiley & Sons, 1950.
Schrieffer, John Robert. *The Theory of Superconductivity*. New York: Benjamin, 1964.
Shoenberg, David. *Superconductivity*. Cambridge, England: Cambridge University Press, 1938.

—*R. Baird Shuman*

Superconductivity at High Temperatures

The Science: K. A. Müller and J. G. Bednorz found a ceramic material that "superconducted" (conducted electricity without resistance) at a temperature much higher than those at which other materials could act as superconductors.

The Scientists:
Karl Alexander Müller (b. 1927), Swiss physicist
Johannes Georg Bednorz (b. 1950), German physicist
Heike Kamerlingh Onnes (1853-1926), Dutch physicist

The Discovery of Superconductivity

At the end of the nineteenth and the beginning of the twentieth century, when modern physics was being created, scientists found that the world did not behave in extreme conditions in the same way that it behaved in ordinary conditions. For example, Albert Einstein discovered that objects moving at speeds close to the speed of light contracted in length. Similarly, Heike Kamerlingh Onnes found that matter behaved in unusual ways at extremely low temperatures. By the first decade of the twentieth century, all gases except helium had been liquefied, and in 1908, using an elaborate device that cooled helium gas by evaporating liquid hydrogen, Kamerlingh Onnes succeeded in liquefying helium. He was then able to use this liquid helium to cool various materials down to temperatures near absolute zero. In 1911, he discovered, to his surprise, that when mercury was immersed in liquid helium and cooled to 4 Kelvins (four degrees above absolute zero), its electrical resistance disappeared. When its temperature rose above 4 Kelvins, mercury lost this property of "superconductivity."

During the decades after Kamerlingh Onnes's discovery, scientists found that other metals and many metal alloys were superconductors when they were cooled to temperatures near absolute zero, but they found no superconductor with a "transition temperature" (the temperature at which superconductivity occurs) higher than 23 Kelvins. They were, however, able to deepen their understanding of superconductivity. For example, in 1933, Karl Wilhelm Meissner, a German physicist, discovered that superconductors expel magnetic fields when cooled below their transition temperatures. This property and the property of resistanceless current flow became the defining characteristics of superconductivity.

Despite these and other discoveries, it was not until 1957 that a satisfactory theory of superconductivity was published. In that year, John Bardeen, Leon N. Cooper, and J. Robert Schrieffer, working at the University of Illinois, used the idea of bound pairs of electrons (Cooper pairs) to explain superconductivity. They showed how the interaction of these electrons with the vibrations of ions in the crystalline lattice of the metal caused the electrons to attract rather than repel one another, and because the movements of neighboring Cooper pairs are coordinated, the electrons could travel unimpeded. Despite these theoretical and experimental advances, however, scientists had been able to achieve only a transition temperature of 23 Kelvins for a niobium-germanium alloy.

The Discovery

Great discoveries are often made by taking risks, and Alex Müller and Georg Bednorz, working at the International Business Machines (IBM) Corporation's laboratory near Zurich, Switzerland, chose to investigate complex metal oxides for superconductivity rather than the usual metals and alloys. Although other scientists had shown that some metal oxides superconducted at very low temperatures, most metal oxides turned out to be insulators (nonconductors). From 1983 to 1985, Bednorz and Müller combined metal oxides to create new compounds to test for superconductivity. More than a hundred compounds turned out to be insulators before Bednorz and Müller heard about a ceramic compound of lanthanum, barium,

Karl Alexander Müller (left) and Johannes Georg Bednorz. (IBM Corporation, AIP Emilio Segrè Visual Archives)

copper, and oxygen that French chemists had made but had failed to test for superconductivity. Bednorz discovered that this compound was indeed a superconductor, and they were able to shift its superconductivity to temperatures as high as 35 Kelvins, twelve degrees above the previous record.

Bednorz and Müller had established that their ceramic material carried electrical current without resistance, but they realized that for a material to be a genuine superconductor, it also had to exhibit the Meissner effect—that is, it had to prevent magnetic fields from entering its interior. In 1986, their tests showed that their material exhibited the Meissner effect. Their results were soon confirmed at several other laboratories, which led to a frantic search among physicists for materials with even higher transition temperatures. When this search led to amazing successes, it became clear that Bednorz and Müller had made a revolutionary breakthrough. In 1987, a little more than a year after their discovery, they received the Nobel Prize in Physics for their work.

IMPACT

After Bednorz and Müller's discovery, other researchers made compounds that superconducted at even higher temperatures, the most famous of which was yttrium-barium-copper oxide, a material that superconducted at almost 100 Kelvins, a temperature higher than that of liquid nitrogen (77 Kelvins). The discovery of this compound's transition temperature led to a flurry of activity among physicists, since it meant that liquid nitrogen—instead of the inconvenient and much more expensive liquid helium—could now be used to study superconductivity. About a year later, researchers at IBM's Almaden Research Center in San Jose, California, announced that they had found a thallium-calcium-barium-copper oxide with a transition temperature of 125 Kelvins.

Because substances with no electrical resistance could have many profitable applications, businessmen, engineers, and government officials were fascinated by these high-temperature superconductors. Corporations and governments invested heavily in their development (about $450 million worldwide in 1990). Politicians and businessmen were convinced of the potential of superconductors in electronics (especially high-speed computers), transportation (especially levitating trains), and power generation and distribution. Several difficulties quickly arose, however, that tempered the initial promise of superconductors. Complex metal oxides are not easily formed into wire, for example, which would be required for many applications. Researchers have tried various ways to solve this problem, but so far they have been unsuccessful. Nevertheless, it seemed almost certain that high-temperature superconductors would eventually transform the way many people live and work.

See also Buckminsterfullerene; Electron Tunneling; Kelvin Temperature Scale; Liquid Helium; Quantized Hall Effect; Superconductivity; Thermodynamics: Third Law.

FURTHER READING

Hazen, Robert. *The Breakthrough: The Race for the Superconductor*. New York: Summit Books, 1988.
Langone, John. *Superconductivity: The New Alchemy*. Chicago: Contemporary Books, 1989.
Mayo, Jonathan L. *Superconductivity: The Threshold of a New Technology*. Blue Ridge Summit, Pa.: TAB Books, 1988.
Prochnow, Dave. *Superconductivity: Experimenting in a New Technology*. Blue Ridge Summit, Pa.: TAB Books, 1989.
Simon, Randy, and Andrew Smith. *Superconductors: Conquering Technology's New Frontier*. New York: Plenum Press, 1988.
                                                                —*Robert J. Paradowski*

Thermodynamics: First and Second Laws

THE SCIENCE: The formulation of the second law of thermodynamics by Rudolf Clausius, along with his insights into the first law of thermodynamics, established the foundation for modern thermodynamics.

THE SCIENTISTS:

Rudolf Julius Emmanuel Clausius (1822-1888), German physicist and mathematician

Sadi Carnot (1796-1832), French physicist

Benoît-Paul-Émile Clapeyron (1799-1864), French engineer

Pierre-Simon Laplace (1749-1827), French physicist and mathematician

Siméon-Denis Poisson (1781-1840), French physicist and mathematician

James Joule (1818-1889), British physicist

Julius von Mayer (1814-1878), German physician and physicist

Hermann von Helmholtz (1821-1894), German physicist

William Thomson, Lord Kelvin (1824-1907), Scottish physicist and mathematician

Ludwig Boltzmann (1844-1906), Austrian physicist

Caloric Theory

The mid-nineteenth century was a time of great interest in thermodynamics, the study of the relationship between heat and other forms of energy. Around 1842, James Joule and Julius von Mayer discovered the first law of thermodynamics, or conservation of energy: Although energy can be changed into different forms, the total energy of an isolated system remains the same. This theory was confirmed by Hermann von Helmholtz in 1847.

In 1850, Rudolf Clausius published a paper in the German *Annalen der Physik* that analyzed the relationship between heat, work, and other thermodynamic variables. Prior to the appearance of Clausius's paper in 1850, the theory of heat, known as the caloric theory, was based on two fundamental premises: the heat in the universe is conserved, and the heat in a material depends on the state of the material. Pierre-Simon Laplace, Siméon Poisson, Sadi Carnot, and Benoît Clapeyron had all developed thermodynamical concepts and relationships that were based upon the assumptions of the caloric theory. By reformulating the first law of thermodynamics using the concept of the internal energy of a system, Clausius showed in his 1850 paper that both assumptions of the caloric theory of heat were incorrect. He

Rudolf Clausius. (Library of Congress)

stated additionally that the natural tendency is for heat to flow from hot bodies to cold bodies and not the reverse. This was the first published statement of what became known as the second law of thermodynamics.

Although the idea seems rather obvious for heat flow that occurs through the process of conduction, the principle stated by Clausius goes much further by asserting that no process whatever can occur that is in conflict with the second law. His 1850 paper was monumental in the development of thermodynamics. It replaced the caloric theory of heat with the first and second laws of thermodynamics, laying the foundation for modern thermodynamics.

EFFICIENCY OF HEAT ENGINES

Between 1850 and 1865, Clausius published an additional eight papers that applied and clarified the second law of thermodynamics. One of his first applications was to the efficiency of a heat engine. A heat engine is any device that absorbs heat from a higher-temperature source, or reservoir, converts part of that energy into useful work, and dumps the rest to a lower-temperature reservoir.

Steam engines are a prime example. In 1824, Carnot had derived an equation for the efficiency of a simple heat engine based strictly on the conservation of energy. In the 1850's, Clausius determined the restrictions on the efficiency of a heat engine by also invoking the second law of thermodynamics in the calculation of efficiency. He showed that the upper limit to the thermal efficiency of any heat engine is always less than one. He concluded that it is impossible to construct any device that will produce no effect other than the transfer of heat from a colder to a hotter body when it operates through a complete cycle. The consequence is that heat energy can not be converted completely into mechanical energy by any heat engine.

IRREVERSIBILITY

Through his applications of the second law to heat engines and other thermodynamic systems, Clausius deduced that processes in nature are irreversible, always proceeding in a certain direction. This is analogous to time only moving forward and not in reverse. Since it is impossible for a heat engine or any other system to convert the heat that it absorbs completely into mechanical work, the system can not return to the same state in which it began. Clausius concluded that the disorder of the system and its surroundings had increased. In a heat engine, for example, the particles that constitute the system are initially sorted into hotter and colder regions of space. This sorting, or ordering, is lost when the system performs work and thermal equilibrium is established.

ENTROPY

In 1865, Clausius published a paper in which he coined the term "entropy" to describe the concept of increasing disorder when processes occur in nature. Becauce the key word in the first law of thermodynamics was "energy," Clausius looked for a similar word to characterize the second law. He finally settled on the word entropy, which originates from a Greek word meaning "transformation." Clausius determined an equation that related entropy to heat and temperature. He then used entropy as a quantitative measure to determine the disorder or randomness of a system.

In his 1865 paper, he restated the second law of thermodynamics in essentially the following form: The change in the entropy of a system interacting with its surroundings always increases. Every event that occurs in the world results in a net increase in entropy. Although energy is con-

---

THE LAWS OF THERMODYNAMICS

FIRST LAW, CONSERVATION OF ENERGY: Energy is neither created nor destroyed; it simply changes from one form to another. Although energy can be changed into different forms, the total energy of an isolated system remains the same. (James Joule and Julius von Mayer)

SECOND LAW, ENTROPY AND IRREVERSIBILITY: Energy available after a chemical reaction is less than that at the beginning of a reaction; energy conversions are not completely efficient. Also, the natural tendency is for heat to flow from hot bodies to cold bodies and not the reverse. (Rudolf Clausius, Sadi Carnot, Lord Kelvin)

THIRD LAW, NERNST HEAT THEOREM: It is impossible to reach a temperature of absolute zero; close to that temperature, matter exhibits no disorder; if one could reach absolute zero, all bodies would have the same entropy, or zero-point energy. (Walther Nernst)

served, an increase in entropy means a reduction in available ordered energy for doing work in the future. Elucidated this way, the second law of thermodynamics is of utmost importance because it imposes practical restrictions on the design and operation of numerous important systems, including gasoline and diesel engines in motorized vehicles, jet engines in airplanes, steam turbines in electric power plants, refrigerators, air conditioners, heat pumps, and the human body.

IMPACT

Because energy conversion is an essential aspect of human technology and of all plant and animal life, thermodynamics is of fundamental importance in the world. The work of Clausius in formulating the second law of thermodynamics laid the framework for modern thermodynamics. The practical significance of his formulation of the second law was recognized on several occasions during his lifetime. He was elected to the Royal Society of London in 1868, received the Huygens Medal in 1870, the Copley Medal in 1879, and the Poncelet Prize in 1883.

Lord Kelvin, who was also instrumental in the development of thermodynamics, pointed out that the principles of heat engines were first correctly established by applying Clausius's second law of thermodynamics and his statement of the first law of thermodynamics. The contributions of Clausius to thermodynamics also formed the basis for future interpretations of the second law of thermodynamics by Ludwig Boltzmann and others in terms of probability, which led to the development of the field of statistical mechanics.

The field of modern thermodynamics and statistical mechanics that evolved from the work of Clausius and other prominent scientists provides immense insights into how the everyday world works, with applications to engineering, biology, meteorology, electronics, and many other disciplines. The operation of engines and the limits on their efficiencies, the operation of refrigerators and the limits on their coefficients of performance and energy efficiency ratings (EER), the function of semiconductors in solid-state circuits as a function of temperature, and the analytical aspects of the human body operating as a thermodynamic engine or fuel cell as it extracts some of the energy released when sugars are metabolized into carbon dioxide and water—all are based on Clausius's formulation of the second law of thermodynamics and his statement of the first law of thermodynamics. Energy-conversion research and the development of alternative energy resources, including solar energy systems, biomass systems, and nuclear power plants, are also dependent on an understanding and application of Clausius's insights into thermodynamic processes and the fundamental laws that govern them.

See also Boyle's Law; Kinetic Theory of Gases; Thermodynamics: Third Law.

Further Reading

Cardwell, D.S.L. *From Watt to Clausius: The Rise of Thermodynamics in the Early Industrial Age*. Ithaca, N.Y.: Cornell University Press, 1971.

Caton, Jerald A. *A Review of Investigations Using the Second Law of Thermodynamics to Study Internal-Combustion Engines*. London: Society of Automotive Engineers, 2000.

Cole, K. C. *The Universe and the Teacup: The Mathematics of Truth and Beauty*. Fort Washington, Pa.: Harvest Books, 1999.

Sandler, Stanley I. *Chemical and Engineering Thermodynamics*. 3d ed. New York: John Wiley & Sons, 1999.

Trefil, James, and Robert M. Hazen. *The Sciences: An Integrated Approach*. 4th ed. New York: John Wiley & Sons, 2004.

—*Alvin K. Benson*

Thermodynamics: Third Law

The Science: Walther Nernst showed that it is impossible to reach a temperature of absolute zero and that, close to that temperature, matter exhibits no disorder. Together, these two statements are known as the third law of thermodynamics.

The Scientists:

Walther Nernst (1864-1941), German physicist
Sir James Dewar (1842-1923), English physicist
Heike Kamerlingh Onnes (1853-1926), Dutch physicist

Reaching Very Low Temperatures

Beginning in the 1870's, scientists were able to achieve temperatures lower than −100° Celsius by allowing gases to expand rapidly. At such low temperatures, the gases themselves often became liquid and could be used to cool other materials to similar temperatures. Work in this area led to the idea of "absolute zero," which was the lowest temperature possible. That temperature was –273° Celsius (0° Celsius is the temperature of an ice-water mixture). During the next thirty years, physicists such as Sir James Dewar tried to see how closely they could approach absolute zero. In trying to reach absolute zero, Dewar invented new equipment for cooling gas and for storing it after it had been liquefied.

Scientists also studied the properties of matter at such very low temperatures, and in doing so they came to two conclusions. First, the closer one approached absolute zero, the more difficult it became to go any lower. This was the result of a combination of practical problems. For example, the colder a sample of matter became, the more rapidly heat leaked into it and warmed it up. Also, at very low temperatures, the liquid gases that were used in the cooling process actually froze into solids, thereby becoming useless in any further cooling. The second conclusion was quite unexpected: At these extremely low temperatures, many properties of matter and energy changed in ways that seemed to contradict existing ideas.

Approaching Absolute Zero

Physicists began asking what led to such dramatic changes, and they also wondered whether it was possible to achieve absolute zero. More than any other physicist of his generation, Walther Nernst was able to provide important answers to these questions. Together, some of these answers became known as the third law of thermodynamics.

To understand the third law of thermodynamics, it is necessary to consider the first two laws. Thermodynamics is the study of a number of forms of energy, including heat energy, mechanical energy, and the energy associated with the orderliness of things. This last kind of energy is called "entropy." More exactly, entropy reflects the degree of disorder that exists; a decrease in orderliness produces an increase in entropy. An example of order is the way in which molecules are arranged in a crystal. An example of disorder is the chaotic way that molecules are scattered in a gas. The first law of thermodynamics states that different forms of energy have to add up: If one kind of energy increases, another kind must decrease. The second law, which is concerned with entropy, states that, in any process, some energy is lost through an increase in disorder. The third law also says something about entropy: As temperature approaches absolute zero, entropy also approaches zero. Thus, at absolute zero, all disorder vanishes.

Walther Nernst was a great experimentalist and an even greater theoretical physicist, and his development of the third law enabled him to combine the two talents. He made ingenious heat measurements at different

Walther Nernst. (Frances Simon, AIP Emilio Segrè Visual Archives)

temperatures close to absolute zero. He also thought deeply about his results and reached a conclusion of great importance. He realized that, at lower and lower temperatures, the total energy of a system becomes more and more nearly equal to its heat energy. In such a case, the total energy is the sum of heat energy and entropic energy. Therefore, if the total energy and heat energy become equal, entropy cannot exist.

This conclusion was astonishing to most of Nernst's contemporaries, who believed that at absolute zero all energy, and not just entropy, became zero. In fact, a certain amount of energy remained at absolute zero. Also, if entropy was zero, then the system was completely ordered. This makes sense, since at zero, even gases would exhibit a solid, crystalized state.

There was also, however, an additional consequence of zero entropy. Nernst realized that, in the process of achieving low temperatures by means of gas expansion, the drop in temperature resulted from converting heat energy into entropy. Therefore, the closer one approached zero (where entropy became zero), the more difficult it became to perform this conversion, and no number of steps could ever reach absolute zero. Thus, the third law makes two statements that seem to be different but are, in fact, closely related: First, as absolute zero is approached, entropy also approaches zero; second, it is impossible to reach absolute zero.

Impact

Nernst's work indicated that quite unexpected things happened at temperatures approaching absolute zero. During the next few years after Nernst's discoveries, additional surprises were in store, many of which were discovered in the laboratory of Kamerlingh Onnes in the Netherlands. For one thing, it was discovered that the electrical resistance of certain metals falls suddenly to zero at a specific temperature, a result that had not been predicted by existing theories. Such "superconductivity" has led to a new understanding of the way in which matter is constructed.

Superconductivity has become important in such practical applications as the efficient transmission of electrical power and the construction of powerful magnets that do not dissipate their power in useless heat. Also, when helium gas is liquefied (at approximately 2° Celsius), it proves to flow much more easily than other "normal" liquids do; this phenomenon is called "superfluidity."

The list of unexpected phenomena that manifest at very low temperatures is a long one, and it will probably grow longer. Some of the most significant advances in physics have come from studies of extremely low temperatures, and Walther Nernst's insights continue to provide important guidance to low-temperature physicists.

See also Boyle's Law; Kinetic Theory of Gases; Thermodynamics: First and Second Laws.

FURTHER READING

Barkan, Diana Kormos. *Walther Nernst and the Transition to Modern Physical Science*. New York: Cambridge University Press, 1999.
Cardwell, Donald Stephen Lowell. *From Watt to Clausius: The Rise of Thermodynamics in the Early Industrial Age*. Ithaca, N.Y.: Cornell University Press, 1971.
Mendelssohn, K. *The World of Walther Nernst: The Rise and Fall of German Science*. London: Macmillan, 1973.

—*John L. Howland*

TROY

THE SCIENCE: Restlessly energetic, ruthlessly self-promoting, wealthy businessman and amateur archaeologist Heinrich Schliemann's excavations of ancient Troy made him a legendary figure.

THE SCIENTISTS:
Heinrich Schliemann (1822-1890), sensational excavator
Frank Calvert (1828-1908), first identifier of site of Troy

A SELF-MADE SELF-PROMOTER

Son of a lowly schoolmaster and clergyman in the north German area of Mecklenburg-Schwerin, Heinrich Schliemann mythologized much of his own life. Apparently his childhood was harsh and his schooling minimal, destining him for a career in trade. He was, however, bright and industrious, a voracious reader with an early flair for languages; he supposedly learned a dozen or so through his years.

He learned business clerking in Amsterdam and at twenty-two joined a major mercantile firm. Sent to Russia, he became a very successful commodities dealer in St. Petersburg. Following family footsteps to California during the gold rush, Schliemann opened a bank there to trade in prospectors' gold. Back in St. Petersburg, Schliemann married a Russian woman, with whom he had three children. During the Crimean War (1853-1856), he profited enormously from dealings in wartime commodities. He made another fortune in cotton and other products during the American Civil War. Several rounds of international travel and the frequenting of museums had stimulated his interest in past cultures. On one tour, he carried off a stone from the Great Wall of China, fascinated by the structure.

HOMERIC STUDIES

Sufficiently wealthy to retire from business by age forty-one, Schliemann hungered to enter some area of scholarly endeavor. In Paris he pursued some formal studies, making intellectual contacts, reading widely, and developing a taste for antiquities. The science of archaeology was still in its infancy, and Schliemann was drawn to it as much as a collector as for scholarly discovery. Beginning a new tour of Mediterranean lands in early 1868, Schliemann studied early archaeological undertakings in Rome and Pompeii. It was the world of early Greece, however, to which he was most attracted.

Like any well-read person of the day, he was deeply familiar with the Homeric epics, and like a good Romantic of his time he was prepared to accept them as factual accounts—even though serious scholars had long rejected them as a pack of legends. Identifying himself with the wandering Odysseus, Schliemann proceeded to the Ionian island of Ithaca, where he made his first primitive venture into some archaeological digging, on what he imagined was the site of Odysseus's palace.

Hungering for new sites and objects, Schliemann stopped in Athens, where a local scholar suggested the Troad at the Dardanelles, the northwestern corner of Asia Minor, in the heartland of the Ottoman Empire (now in Turkey). Schliemann was directed to the hill of Pinarbashi (Bunarbashi), which some antiquarians thought was the site of ancient Troy, and he was advised to consult a local expert, the Englishman Frank Calvert, then American vice-consul for the region. Making his way there in

August of 1868, Schliemann initially avoided Calvert, reconnoitering and then undertaking some ill-defined and fruitless excavations at Pinarbashi. Only when about to leave did he meet Calvert. The latter had spent years exploring the area's topography and sites. First interested in Pinarbashi, he had come to reject it as the site of Troy, which he now firmly believed was the hill called Hisarlik. He had even purchased a portion of the hill and wanted to excavate it himself but lacked financial means.

HISARLIK: UNDISCIPLINED EXCAVATIONS

Calvert gave Schliemann a crash-course in what Hisarlik represented. Calvert recognized that this wealthy enthusiast had the means to do what he himself could not afford to do, while Schliemann recognized a golden opportunity at hand. In proposing a partnership with Schliemann, however, Calvert not only shared his dream but sacrificed it to an opportunist whose character he had not understood.

While months were spent in correspondence and securing permissions from the local Turkish authorities, Schliemann began consolidating his standing. He published a book exaggerating his work on Ithaca but staking his claims as a serious archaeologist. With that he secured an honorary doctorate from the University of Rostock, thus acquiring an instant scholarly stature that Calvert, the gentleman-antiquarian, lacked. Moreover, in a quick trip to the United States, Schliemann took out American citizenship, which he used deviously to obtain a divorce from the Russian wife who had refused to follow him in his adventures. Thus freed to extend his philhellenism, Schliemann found himself a new Athenian bride in Sophia Engastromenou, all of seventeen years old to his forty-seven.

In April, 1870, Schliemann began serious excavations at Hisarlik. From the start, he engaged in constant duplicity, breaking agreements with Calvert and practicing forms of digging that were clear vandalism by modern archaeological standards. Ignoring Calvert's advice, he had large

The Mask of Agamemnon, from one of the Trojan excavations.

trenches dug, culminating in a huge north-south gash across the hill. The successive campaign years turned up numerous finds that reinforced the identification with Troy. However, Schliemann was quite unprepared for the complex layering of strata in his quest to identify the Troy of Homer's King Priam. Further conflicts developed over Schliemann's cheating Calvert out of his share of treasures, and there was even a rupture between them when Calvert argued in print against reckless interpretations of the site that Schliemann was circulating in his orgy of self-serving publicity.

PRIAM'S TREASURE

The climax of Schliemann's excavations came on May 31, 1873, when Schliemann came upon a body of copper and gold objects, including jewelry. This trove he proclaimed Priam's treasure, reporting that it was recovered with the help of his wife. In fact, Sophia was back in Athens at the time. Some critics have even speculated that the "treasure" was a plant— objects Schliemann had purchased on the black market and sequestered on the site for "discovery." Defying his contract with the Turkish government, Schliemann smuggled these objects out to Athens, to install them in his home there for exhibition and photography. A picture of his wife bedecked in "Helen's Jewels" was circulated worldwide as the peak of Schliemann's self-promotion. These treasures eventually found their way to Berlin, from which they were carried off into obscurity by the Russians after World War II; in 1993 it was finally revealed that they were preserved at Moscow's Pushkin Museum.

Schliemann faced fury and long legal actions from Constantinople, and only after a financial settlement was he allowed further access to Troy. In 1874, Schliemann published in book form his excavation reports, in which he consolidated his fame as the discoverer of Homer's Troy, in the process burying any credit due Calvert. Indeed, in his autobiographical writings, Schliemann even appropriated from Calvert the story that he had nourished since childhood about his determination to find and reveal Homer's Troy.

On the basis of his sensational work at Troy, Schliemann was allowed to conduct excavations in Greece at Mycenae in 1876, where he cleared the grave circle and discovered its famous burial masks. In 1878-1879, after an uneasy reconciliation with Calvert, Schliemann pursued new excavations at Troy. He also ventured some further "Homeric" explorations at Ithaca (1878), Orchomenos (1881), and Tiryns (1884-1885), while continuing his prolific outpouring of writings and publications. A celebrity of worldwide standing and now one of the great men of Greece, he built a grand mansion in downtown Athens for himself and Sophia; it still stands.

Still, the perplexities of Troy drew him, and, with Calvert's collaboration, Schliemann undertook new explorations of the area in 1882. He continued involvement with the site, attending an international conference held there in 1889 to clarify the identity of Hisarlik. His plans for further investigations in 1890 were cut short unexpectedly by his death during a visit to Naples. Schliemann's remains were brought back to Athens and buried in a grandiose neoclassical mausoleum on a hilltop in the city's main cemetery.

IMPACT

The tangled explication of the various layers of Hisarlik's settlements was resumed by Schliemann's assistant, Wilhelm Dörpfeld, and continue to the present day, all on a more scientific scale. Schliemann did demonstrate that the stories of the Trojan War corresponded to tangible evidence and that Hisarlik was the site of the ancient Troy, but his brutal excavation techniques ironically destroyed much of what remained of the city of the Trojan War. Acclaimed as "the father of archaeology," Schliemann awakened a broad public to this new science, but his methods now evoke horror, while his shameful suppression of Calvert's role is now evident.

See also Dead Sea Scrolls; Pompeii; Rosetta Stone; Stonehenge.

FURTHER READING

Allen, Susan Heuck. *Finding the Walls of Troy: Frank Calvert and Heinrich Schliemann at Hisarlik*. Berkeley: University of California Press, 1999.
Calder, William M., III, and David A. Traill, eds. *Myth, Scandal, and History: The Heinrich Schliemann Controversy*. Detroit: Wayne State University Press, 1986.
Ludwig, Emil. *Schliemann: The Story of a Goldseeker*. Translated by D. F. Tait. Boston: Little, Brown, 1931.
Schliemann, Heinrich. *Troy and Its Remains: A Narrative of Researches and Discoveries Made on the Site of Ilium and in the Trojan Plain*. London: J. Murray, 1875.
Traill, David A. *Schliemann of Troy: Treasure and Deceit*. London: J. Murray, 1995.
Wood, Michael. *In Search of the Trojan War*. Berkeley: University of California Press, 1988.

—*John W. Barker*

Uniformitarianism

THE SCIENCE: James Hutton's "dynamic equilibrium"—the theory that Earth's formation was the result of a cyclic process of erosion and uplift which, in turn, was the result of the compounding of the ordinary action of water and heat in geologic time—laid the foundation for Charles Lyell's uniformitarian geology and established the "deep time" necessary to Charles Darwin's evolutionary theory.

THE SCIENTISTS:

James Hutton (1726-1797), a Scottish natural philosopher

John Playfair (1748-1819), a Scottish geologist and mathematician, and chief popularizer of Hutton's theory

Sir James Hall (1761-1832), Scottish geologist and chemist

Abraham Gottlob Werner (1750-1817), a professor at Freiburg Mining Academy

George-Louis Leclerc, Comte de Buffon (1707-1788), a French naturalist

HUTTON'S DURABLE EARTH

On March 7, 1785, members of the Royal Society of Edinburgh assembled to hear a much-anticipated paper. Its author, James Hutton, was ill that day and had chosen his closest friend, the renowned philosophical chemist Joseph Black, to deliver the first part of a new theory in a work titled "Concerning the System of the Earth, Its Durability and Stability." Four weeks later, on April 4, Hutton had recovered sufficiently to present the second part of his theory. In July, he privately printed and circulated an abstract of the paper. Eventually, in 1788, the full ninety-five-page manuscript from which Hutton's papers had been drawn was published in the first volume of the *Transactions* of the Royal Society as "Theory of the Earth: Or, An Investigation of the Laws Observable in the Composition, Dissolution, and Restoration of Land upon the Globe."

Hutton's paper was written in the context of a well-established consensus in eighteenth century geological science. The French naturalist Georges-Louis Leclerc, the comte de Buffon, had provided a general framework for the consensus in the initial volumes of his *Histoire naturelle, générale et particulière* (1749-1789; *Natural History, General and Particular*, 1781-1812), and again in his *Époques de la nature* (1778; the epochs of nature). According to Buffon, the Earth originated as solar matter. As Earth cooled, a universal ocean covered its surface. Sedimentation of materials suspended in this primitive ocean produced rock strata, which were exposed as the ocean receded.

In the 1780's, Abraham Gottlob Werner, a professor at the Freiburg Mining Academy in Saxony, supplemented Buffon's cosmogony with a stratigraphy that distinguished four mineralogical groups according to the order in which they had settled out of the universal ocean. Most primitive were chemical precipitates such as granite. Settling out next, at lower elevations, were heavier materials such as limestone, followed by basalts at still lower elevations and sand and other alluvial deposits. Werner's stratigraphy would potentially accommodate a great variety of geological phenomena and gain widespread acceptance. Indeed, when professor of natural history John Walker offered the first series of lectures in geology at the University of Edinburgh in 1781, it was Werner's stratigraphy that he introduced.

Plutonism Joins Neptunism

What the audience for Hutton's Royal Society paper heard was an audacious departure from the Wernerian consensus. To be sure, Hutton's theory acknowledged that water was one of the primary agents in geological change. During the 1750's and 1760's, Hutton had been a highly innovative agricultural improver on his Berwickshire estate and knew well the power of erosion. However, for Hutton, water's geological effect was destructive. The action of water could not explain dramatic features of the Earth's topography such as mountain ranges, nor could it adequately explain the presence of unconformities in rock strata.

For these phenomena, Hutton required a more constructive force. A graduate of the University of Edinburgh in 1743, Hutton had had a longtime interest in chemistry and, during the 1740's, he had even invented an improved method of producing the ammonium chloride used in soldering metals. In 1768, Hutton leased his farm and returned to Edinburgh. It was at this time that Hutton developed his friendship with Joseph Black, a pioneer in the study of heat, and Black's former student, James Watt, the inventor of the modern steam engine. These friendships soon suggested to Hutton a second agency in geological change. Subterranean heat, he began to argue, drove the terrestrial machine.

Hutton was combining the so-called Neptunian consensus with a new Plutonism: As does the sea, the underworld could rise and fall. The origins of this conception of dynamic equilibrium again lay in Hutton's own past. As an undergraduate at the University of Edinburgh, Hutton had studied with Colin Maclaurin, one of the eighteenth century's most effective popularizers of Sir Isaac Newton's ideas. In the autumn of 1744, Hutton had begun to study medicine at Edinburgh; in 1747, he had transferred to the University of Paris; and, in 1748, he had enrolled in the center of medical Newtonianism at the University of Leiden, where he completed his medical degree with *Dissertatio physico-medica inauguralis de sanguine et circulatione microcosmi* (1749; *James Hutton's Medical Dissertation*, 1980).

Just as a balance of centrifugal and gravitational forces produced the solar system in Newtonian astronomy, and just as systolic and diastolic forces circulate the blood (or just as the piston pushes and pulls in Watt's double-acting steam engine), uplift complemented erosion in Hutton's system of the Earth. As water dissolved rock, the products of sedimentation accumulated, generating intense heat and pressure. The intense heat and pressure, in turn, caused rapid expansion and uplift. Even as old continents become new oceans, old oceans become new continents.

Dynamic Equilibrium = Uniformitarianism

Hutton's argument had profound implications for eighteenth century geology. Whereas Buffon conceived the history of the Earth proceeding in one direction of greater cooling and erosion, Hutton did for matter in time what Newton had done for matter in space and conceived the laws of motion in terms of reciprocal action and reaction. Whereas Neptunism implied catastrophism, the assumption that the geological processes of the past were of a qualitatively greater magnitude than the geological processes of the present, Hutton's dynamic equilibrium implied a uniformitarianism, in which the observed geological processes of the present were the key to understanding the geological processes of the past. Also, whereas catastrophism offered a way to reconcile biblical accounts of the Creation and the Flood, Hutton entirely ignored Genesis as a source of geological knowledge.

"Deep" Time

Finally, Hutton undercut Werner's stratigraphy. In the years after 1785, Hutton traveled to sites across Scotland—most famously, to the unconformity at Siccar Point—to gather evidence for his theory. This evidence indicated that granite had not originated as an aqueous precipitate but had crystallized from molten magmas. It also revealed that granite, which Werner had made the oldest rock, was in some cases intruded upward into sedimentary strata. Conversely, the evidence also showed the ancient volcanic origin of basalts. By recognizing the igneous origin of many rocks, Hutton reconfirmed that the forces of geological change acted uniformly in past and present. He also reconfirmed that geological processes did not work in one direction and, therefore, no inherent limit could be set regarding Earth's geological history. In short, Hutton had discovered deep time. "We find," Hutton famously concluded in his 1788 paper, "no vestige of a beginning, —no prospect of an end."

IMPACT

James Hutton extended the Scottish tradition of conjectural history to the economy of nature. From the perspective of deep time, cycles of dissolution and composition succeeded one another in a spontaneous and, ultimately, benevolent order. As if by an invisible hand, the nearly imperceptible action of heat and water shaped the grandest features of the physical environment. Still, for all the assurance of durability and stability, Hutton's theory faced intense criticism in the years after 1788. The specter of the French Revolution rendered Hutton's defense of design an invitation to atheism, his understanding of nature's uniformity an incitement to political turbulence.

In 1795, a very ill Hutton attempted to rebut these charges in the two sprawling volumes. Most responsible for keeping Huttonian geology before a scientific public, however, were his two companions on the trip to Siccar Point in 1788, the chemist and geologist James Hall and the mathematician John Playfair. Under the pressure of criticism, Hutton's geology underwent its own reconstruction. In a series of ingenious experiments between 1798 and 1805, Hall proved aspects of Hutton's theory, demonstrating, for example, the thermal metamorphosis of limestone into marble. In 1802, geologist Playfair published *Illustrations of the Huttonian Theory of the Earth*. Beautifully presented in compelling prose, Playfair's illustrations were no longer organized as mere proofs of an a priori theory but as the empirical foundation for a historical narrative.

It was Hall's experimental and Playfair's empirical Huttonianism, that, in 1830, provided the precedent for the scientific uniformitarianism of British geologist Charles Lyell's *Principles of Geology* (1830-1833). It was the same Huttonianism that, in 1859, provided two crucial elements to the variational evolution of Charles Darwin's *On the Origin of Species by Means of Natural Selection*: the notion that evolutionary change could take place by an incremental accumulation of small variations and the discovery of time that allowed this slow agency to work.

See also Continental Drift; Earth's Core; Earth's Structure; Fossils; Geologic Change; Geomagnetic Reversals; Hydrothermal Vents; Mass Extinctions; Microfossils; Plate Tectonics; Radiometric Dating; Seafloor Spreading.

FURTHER READING

Dean, Dennis R. *James Hutton and the History of Geology*. Ithaca, N.Y.: Cornell University Press, 1992.

Gould, Stephen Jay. *Time's Arrow, Time's Cycle: Myth and Metaphor in the Discovery of Geological Time*. Cambridge, Mass.: Harvard University Press, 1987.

Hutton, James. *System of the Earth, 1785, Theory of the Earth, 1788, Observations on Granite, 1794, Together with Playfair's Biography of Hutton*. Darien, Conn.: Hafner, 1970.

McIntyre, Donald B., and Alan McKirdy. *James Hutton: The Founder of Modern Geology*. Edinburgh: Stationery Office, 1997.

Repcheck, Jack. *The Man Who Found Time: James Hutton and the Discovery of the Earth's Antiquity*. Reading, Mass.: Perseus Books, 2003.

—*Charles R. Sullivan*

VAN ALLEN RADIATION BELTS

THE SCIENCE: James Van Allen pioneered the use of artificial satellites for Earth studies, which led to the discovery of electrically charged particles trapped within the Earth's magnetic field.

THE SCIENTISTS:
James Van Allen (b. 1914), American physicist and naval officer

REACHING TOWARD SPACE

Until the late 1950's, studies of the Earth's magnetic field could be performed only from the Earth's surface. This hindered the understanding of how the field is generated, as well as the determination of the field's shape and strength and the volume of space that it occupies. This changed, however, with the development of artificial satellites. The V-2 rockets used by the Germans during World War II (1939-1945) to destroy English cities could reach an altitude of 100 kilometers. Although they did not reach the speeds necessary to place a satellite in orbit, they were a step in the right direction. After the war, captured German scientists and rockets provided the basis for the United States' space efforts.

James Van Allen started studying cosmic rays as an undergraduate. He received his Ph.D. in 1939. During the war, he served as a naval officer and worked on a proximity fuse for artillery shells. This device used a radar signal that emanated from the shell once the shell had reached its target, triggering the fuse that caused the shell to explode. Van Allen worked to miniaturize the electronic components needed for the small confines of the shell.

After the war, Van Allen worked to reduce the instrument packages being sent aloft in the captured V-2 rockets. These rockets could go higher because they were now lifting smaller payloads, but they were still not capable of placing these payloads into permanent Earth orbit. By 1954, however, Van Allen and his colleagues began talking about the possibility of using the larger, more powerful rockets that were then under development. In 1955, President Dwight D. Eisenhower announced that the United States would launch an artificial satellite within two years.

Scientists designated the time period from July 1, 1957, to December 31, 1958, as the International Geophysical Year. During this time period, the Earth and its surrounding area were to be studied intensely by scientists around the world to learn more about the planet. The Soviets announced their intention to launch an artificial satellite as part of this study. On October 4, 1957, they launched Sputnik 1 into Earth orbit.

Beyond the Limits

Van Allen's war experience proved to be invaluable in reducing the weight of experiments launched by the United States' less powerful rockets. Although the payloads were smaller, they were more sophisticated because of the efforts of Van Allen and others. On July 26, 1958, the United States included a Geiger-Müller counter in its launch of the Explorer 4 satellite to detect space radiation. When the counter's radio signal was transmitted to Earth for analysis, it did something strange: It increased to a maximum, decreased to zero, and then increased again to maximum. Van Allen correctly interpreted this not as a result of an actual decrease in radiation but as a result of the instrument's inability to handle high levels of radiation. This is analogous to the distortion one hears when the volume of a radio is turned too high, driving the electronics beyond their design limits.

Further study revealed the nature of the radiation. The Earth's magnetic field temporarily traps electrons and other electrically charged particles emitted by the Sun. These particles constantly flow from the Sun and are known as the solar wind. Some of the particles also may come from Earth's upper atmosphere as its gases interact with the solar particles. The Earth's magnetic field fans out at the magnetic pole in the Southern Hemisphere, arcs over Earth's equator, and converges on the magnetic pole in the

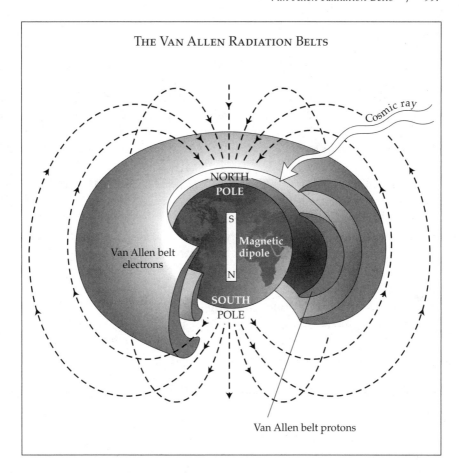

THE VAN ALLEN RADIATION BELTS

Cosmic ray

NORTH POLE

S

Magnetic dipole

N

Van Allen belt electrons

SOUTH POLE

Van Allen belt protons

Northern Hemisphere. The field is strongest at the poles and weakest halfway between them. Particles such as electrons enter the field and spiral along the field lines. As the field strength increases near the poles, the particles bounce off this area and spiral toward the opposite poles. The particles may perform this spiral bounce motion many times before escaping into outer space.

There are two broad bands, or "belts," in the Earth's magnetic field that have high radiation levels. Aligned with the center of Earth, they are both doughnut-shaped, with crescent-shaped cross sections. The inner belt begins at 3,000 kilometers above the Earth's surface and is at its thickest portion at 5,000 kilometers. The outer belt is 16,000 kilometers from the Earth's surface and is 6,500 kilometers thick. Although the electrically charged particles consist mostly of electrons, the inner belt does contain some protons and other particles. In honor of Van Allen's discovery, these belts of high radiation were named the Van Allen radiation belts.

Impact

Orbiting satellites provide a very convenient method for communication, navigation, and Earth monitoring. As more satellites are placed in orbit, however, convenient orbital paths are filling up, creating the need to expand space usage farther from Earth. The Van Allen radiation belts complicate this goal because their high levels of radiation require the use of extra shielding for the satellite's instruments, thereby increasing the mass of the payload.

The spiraling and bouncing behavior of the particles in the field may provide an idea for a solution to the energy crisis. Because convenient energy sources are being depleted, other methods must be found for generating the needed energy. One possibility is the fusion of hydrogen nuclei into helium, a process that releases huge amounts of thermonuclear energy. For fusion to occur, however, high temperatures of millions of degrees are needed. No material known is capable of containing such heat. It is, however, possible to produce a magnetic "bottle," in which the high-temperature hydrogen plasma spirals along the field and bounces at the ends of the "bottle," where the field increases in strength. The idea for such a device was taken from analysis of the Van Allen radiation belts. The magnetic bottle is especially useful in a controlled thermonuclear reaction. This is one example of how discoveries in one field of science may influence work in other areas of science, sometimes to the benefit of all of humanity.

See also Geomagnetic Reversals; Solar Wind.

Further Reading

Van Allen, James A. *Origins of Magnetospheric Physics*. Expanded ed. 1983. Reprint. Washington, D.C.: Smithsonian Institution Press, 2004.
Walt, Martin. *Introduction to Geomagnetically Trapped Radiation*. New York: Cambridge University Press, 1994.

—*Stephen J. Shulik*

Very Long Baseline Interferometry

The Science: Very long baseline interferometry made it possible to study distant regions of the universe that had not been studied before.

THE SCIENTISTS:
Sir Martin Ryle (1918-1984), radio astronomer
Karl Jansky (1905-1950), pioneer radio astronomer

LIMITS OF THE RADIO TELESCOPE

Astronomers have always wanted to study stars, planets, quasars, and other objects that are too far from Earth to be seen with conventional optical telescopes. Scientists discovered that they could learn about distant objects by studying the waves of radiation that those objects emit. They constructed radio telescopes—long, dishlike apparatuses that faced up to the sky and that could collect incoming radiation waves. By studying those waves, astronomers could "see" into distant areas of the universe.

Radio astronomy, which uses radio telescopes to study objects and phenomena in the universe, was "discovered" by Karl Jansky in 1929. At that time, he was working at Bell Telephone Laboratories, studying a strange hiss that could be heard on telephone lines. Using the information he collected with a very crude radio telescope, he discovered that the hiss was caused by radio signals that came from the Milky Way.

After World War II ended in 1945, scientists built bigger and bigger radio telescopes. The largest radio telescope, which was built in Arecibo, Puerto Rico, had a 1,000-foot (300-meter) reflector dish to collect radio waves. Radio telescopes must be large because radio waves, which are a kind of radiation, are much longer than the waves of other kinds of radiation, such as visible light. An ordinary optical telescope collects waves of visible light, which are measured in microns (millionths of a meter). Radio waves, however, can be centimeters (hundredths of a meter) or even meters long. Even the largest single-dish radio telescopes are less accurate than the much smaller optical telescopes are, because the waves they collect are so much longer.

An interferometer is an instrument that studies an object by comparing sets of waves, such as radio waves, that are emitted by that object. Two or more separate radio telescopes can be combined by being connected with a cable to "create" a very large and accurate radio telescope. The large radio telescope that is created in this way is an interferometer, because it can compare the sets of waves collected by the separate radio telescopes. This means that scientists can use two radio telescopes to create a telescope to do the same job that it would take a single, huge telescope to do. The distance between the two original telescopes is called the baseline, so the science of creating and using interferometers in this way is called baseline interferometry.

A World-Spanning Radio Telescope

Pioneering scientists such as Martin Ryle, whose work in interferometry earned for him the 1974 Nobel Prize in Physics, wanted to make interferometers that would be even larger, and therefore more accurate, than those that could be made by connecting radio telescopes with a cable. Ryle's development of a technique called "aperture synthesis" made it possible to increase dramatically the accuracy of connected radio telescopes. It was not until the atomic clock was perfected, however, that even larger interferometers could be constructed and the techniques of very long baseline interferometry (VLBI) came into existence.

Using the extremely accurate atomic clock, astronomers at observatories very far from each other could monitor the same radio source simultaneously and record the resulting information on tape, along with the signals from the atomic clock. Later, the clock signals could be used to synchronize the information from the two radio telescopes. The result would be the same as if the two telescopes had been connected physically while they were monitoring the radio source. It was no longer necessary for telescopes to be in direct contact while they were monitoring a radio source.

By using the atomic clock, astronomers can make interferometers of two or more radio telescopes that are many kilometers apart. The maximum baseline that can be used for VLBI is the diameter of the Earth: 12,700 kilometers. With a baseline so large, it is possible to obtain extremely accurate results. In fact, the accuracy, or resolution, of VLBI observations is the same as seeing a pinhead at a distance of 200 kilometers.

Although VLBI technology has provided scientists with very precise information, its techniques are difficult and time consuming to use. For example, when radio telescopes that are very far apart are combined, they must collect signals for a long period of time in order to obtain a high-resolution radio signal. Also, two radio telescopes that are separated by the diameter of the Earth form only two tiny parts of the collector that they are trying to create. They cannot form a proper image, although they can determine whether a radio source does contain much detail. More radio telescopes are needed to create a more complete VLBI collector. The results from six radio telescopes, for example, can produce the most detailed astronomical images that can be formed at any wavelength.

Impact

VLBI technology has greatly increased the resolution of radio telescopes. Before VLBI existed, radio astronomy had relatively poor resolution. The increased resolution made possible by VLBI gave astronomers an extremely detailed view of the universe.

With VLBI technology, astronomers can explore objects that were previously too far away to be examined. One example of such objects is quasars, which are quasi-stellar objects that are a source of both visible and radio waves. Quasars are almost starlike in appearance, but they are among the most powerful and most distant sources of energy in the universe. High-resolution VLBI observations have revealed that small, jetlike structures shoot outward from the active regions of many quasars. These "small" jets of matter are only a few light-years long (one light-year is about 9.46 billion kilometers); they are aligned with much larger jets of matter that are often millions of light-years long. The existence and position of these jets of matter indicate that there is some connection between the jets and the quasar cores in which they are probably generated.

The development of VLBI technology promises to give astronomers an even more detailed image of the universe. With this technology, scientists are likely to learn more about both quasars and the mysterious "black holes" (extremely dense and invisible stellar objects) that may be the source of their power.

See also Cosmic Microwave Background Radiation; Ionosphere; Isotopes; Pulsars; Quasars; Radio Astronomy; Radio Galaxies; Radio Maps of the Universe; Wilkinson Microwave Anisotropy Probe.

FURTHER READING

Bracewell, Ronald N. *The Fourier Transform and Its Applications.* 3d rev. ed. New York: McGraw-Hill, 1987.
Brown, Robert Hanbury, and A. C. B. Lovell. *The Exploration of Space by Radio.* New York: John Wiley & Sons, 1958.
Dudgeon, Dan E., and Russell M. Mersereau. *Multidimensional Digital Signal Processing.* Englewood Cliffs, N.J.: Prentice-Hall, 1984.
Kellerman, Kenneth L., and A. Richard Thompson. "The Very-Long Baseline Array." *Scientific American* 258 (January, 1988): 54-61.
Mills, B. Y. "Apparent Angular Sizes of Discrete Radio Sources." *Nature* 170 (1952): 1063-1065.
Thompson, A. R., James M. Moran, and George W. Swenson, Jr. *Interferometry and Synthesis in Radio Astronomy.* New York: Wiley, 1986.
Verschuur, G. L. *The Invisible Universe: The Story of Radio Astronomy.* Berlin: Springer-Verlag, 1974.

—*David R. Teske*

VIRUSES

THE SCIENCE: The Dutch bacteriologist Martinus Beijerinck demonstrated that cell-free extracts prepared from plants with tobacco mosaic disease could transmit the infection to healthy plants. The preparation contained what Beijerinck called *contagium vivum fluidum*, infectious material that would only replicate in living tissue, and represented the first evidence for the existence of what became known as viruses.

THE SCIENTISTS:
Martinus Beijerinck (1851-1931), Dutch bacteriologist who demonstrated that viruses require living tissue for replication
Adolf Mayer (1843-1942), Dutch biologist who first demonstrated transmission of a viral disease
Dimitri Ivanovski (1864-1920), Russian biologist who first associated a filterable agent with tobacco mosaic disease
Friedrich Löffler (1852-1915), professor of hygiene who codiscovered the foot-and-mouth disease virus
Paul Frosch (1860-1928), codiscoverer, with Löffler, of the foot-and-mouth disease virus

BEYOND THE GERM THEORY

During the last decades of the nineteenth century, Robert Koch and others discovered that bacteria represented the agents behind many human illnesses, which led to the "germ theory of disease," the idea that most illnesses are caused by bacteria. The ability to grow these organisms on laboratory media played a major role in the formation of Koch's postulates, a series of experimental steps linking a disease with a specific organism.

The growing list of diseases found to be associated with bacterial infections gave rise to a belief that most diseases are the result of infection by such microscopic agents. Vaccines had been developed by the 1880's against what are now known to be virally induced diseases, most notably against smallpox and rabies. However, the reluctance to carry out infection in humans by applying extracts from infected tissues meant that the soluble nature of these agents was overlooked; Koch's postulates could not be applied. The difficulty in growing many viral agents in the laboratory, as well as lack of animal models, would remain a problem in applying Koch's postulates to viral diseases well into the twentieth century.

The first experimental transmission of a viral disease could arguably be attributed to Adolf Mayer, director of the agricultural station at Wageningen in The Netherlands. Mayer had been studying tobacco mosaic disease (TMD), a name he coined, which was having a significant economic impact on tobacco growers. Mayer demonstrated that one could transmit the disease to healthy plants by spraying the sap extracted from diseased plants. He attempted to link bacterial agents he had isolated from the diseased plant with TMD by applying Koch's postulates, but once again the inability to culture any specific organism made this impossible. Since it appeared the agent was removed from the preparation following the filtration step, Mayer incorrectly suggested that the agent was probably a bacterium.

The Russian biologist Dimitri Ivanovski repeated and extended Mayer's work in 1892. Mayer had used a double layer of filter paper to remove any bacteria or other cells from his preparation. Instead of using filter paper, however, Ivanovski prepared cell-free filtrates from diseased tobacco plants while using newly developed porcelain Chamberland filter-candles. He was able to transmit the disease to healthy plants even in the absence of bacteria. Ivanovski's conclusion was that a toxin was probably associated with the disease, reflecting the recent discovery by Emil Behring of the relationship between a toxin and human diphtheria. As late as 1903, Ivanovski maintained that the agent behind TMD was probably a bacterium that could not be cultured.

A Contagious Living Fluid

Martinus Beijerinck was probably unaware of Ivanovski's earlier work on the nature of the tobacco mosaic disease agent. In 1898, Beijerinck was collaborating with Mayer in the study of the disease, and he repeated the filtration experiments that, unknown to him, were first conducted six years earlier by Ivanovski. Beijerinck's conclusion was that the sap contained a *contagium vivum fluidum*, a contagious living fluid.

In a more detailed analysis of the agent, Beijerinck first demonstrated it would not grow on the culture media generally used to grow or maintain bacteria. Nor would the agent grow in the sap itself. Beijerinck concluded the agent could not be a bacterium. Further, he found the agent was capable of diffusing through agar, indicating that it was a soluble substance and that it was stable over a period of months even when dried. His work was reported in a publication later that same year.

Beijerinck also carried out studies on the development of tobacco mosaic disease itself. He observed the agent spread through the plant through the phloem, and he noted that it had a preference for young, growing leaves. By passing the sap from plant to plant, Beijerinck demonstrated that it was capable of reproduction—unlike a toxin, which would have lost viability as it became diluted. Beijerinck's conclusion—that the TMD agent was neither bacterial in nature nor a toxin, but that it required living tissue in which to reproduce—set his work apart from that carried out earlier by Ivanovski. Beijerinck correctly has been given priority in the discovery of viruses.

VIRUSES NEED LIVING TISSUE

Shortly after the work by Beijerinck on TMD, the first demonstration of a disease in animals that could be transmitted by cell-free extracts was achieved. Friedrich Löffler, head of a Prussian Research Commission for the study of foot-and-mouth disease, and his collaborator Paul Frosch, a colleague of Robert Koch at Koch's Institute of Infectious Diseases in Berlin, transmitted the disease using extracts from vesicles isolated from infected cattle. Clearly, agents too small to be observed with standard microscopes, and which required living tissue in which to replicate, were associated with diseases.

Tobacco mosaic virus went on to play a major role in the nascent field of virology. It was the first virus to be purified free from host tissue (1935). Also—unlike most organisms, in which DNA was determined by the 1940's to be the genetic material—TMV was the first agent found to contain RNA as a genome.

IMPACT

Though Beijerinck was not the first to observe the role for a filterable agent as an etiological agent for (plant) disease, he demonstrated that the agent could not be grown in culture media, which likely meant it was not a bacterium. Further, the fact the filterable agent could be shown to multiply eliminated the possibility of its being a toxin. Beijerinck's definition of the *contagium vivum fluidum*, however, could not imply an understanding of a "virus" in the same context as would later be determined. The modern concept of a virus required a leap in understanding that was premature for the science of the day.

Beijerinck's discoveries were particularly significant in that he demonstrated that the agent required living tissue in which to reproduce. Shortly afterward, Löffler and Frosch reported that an analogous agent was associated with foot-and-mouth disease in animals. This led to a twenty-five-year debate over the nature of such "viruses": Were they particles or enzymes? The question was settled only with the independent codiscovery of bacterial viruses by Frederick Twort and Felix d-Herelle, as well as the development of the electron microscope, which allowed viruses to be visualized.

During these same decades, filterable agents were also demonstrated to be etiological agents of human or other animal diseases such as yellow fever, polio, rabies, and possibly even cancer. None of these "organisms" could be grown on laboratory media; they could only be shown to replicate in the animal itself. Scientists gradually came to the conclusion that viruses represented a form of "life" that could not be considered as either animal or plant.

See also AIDS; Human Immunodeficiency Virus; Hybridomas; Immunology; Oncogenes; Polio Vaccine: Sabin; Polio Vaccine: Salk; Smallpox Vaccination; Yellow Fever Vaccine.

FURTHER READING

Dimmock, N. J., A. J. Easton, and K. N. Leppard. *Introduction to Modern Virology*. 5th ed. Malden, Mass.: Blackwell Science, 2001.

Fraenkel-Conrat, Heinz. "The History of Tobacco Mosaic Virus and the Evolution of Molecular Biology." In *The Plant Viruses*, edited by M. H. V. van Regenmortel and Heinz Fraenkel-Conrat. New York: Plenum Press, 1986.

Helvoort, Ton van. "When Did Virology Start?" *ASM News* 62 (1996): 142-145.

Knipe, David, Peter Howley, and Diane Griffin. *Fields' Virology*. 2 vols. 4th ed. New York: Lippincott Williams and Wilkins, 2001.

Rott, Rudolf, and Stuart Siddell. "One Hundred Years of Animal Virology." *Journal of General Virology* 79 (1998): 2871-2874.

Zaitlin, Milton. "The Discovery of the Causal Agent of the Tobacco Mosaic Disease." In *Discoveries in Plant Biology*, edited by S. Kung and S. Yang. Hong Kong: World, 1998.

—Richard Adler

VITAMIN C

THE SCIENCE: Albert Szent-Györgyi isolated "hexuronic acid," a substance that years later was proven to be vitamin C.

THE SCIENTISTS:
Albert Szent-Györgyi (1893-1986), Hungarian biochemist who won the
    Nobel Prize in Physiology or Medicine in 1937
Charles Glen King (1896-1988), American biochemist and nutritionist
Joseph L. Svirbely (1906-1995), American chemist

ORANGES AND OXEN

Vitamin C is both a complex chemical substance and the physiological linchpin in the deficiency disease scurvy. Physicians in the Middle Ages had recognized some aspects of this disease, which was characterized by weakness, swollen joints, a tendency to bruise easily, bleeding from the gums, and the loss of teeth. It was not until the eighteenth century, however, that the Scottish physician James Lind recognized that these symptoms constitute a disorder caused by defective nutrition. His experiments on sailors during long ocean voyages showed that the ingestion of certain fruits and vegetables could cure the disease.

The most significant step toward the discovery of vitamin C was made in 1907, when Axel Holst, a bacteriologist, and Theodor Frölich, a pediatrician, published their discovery that, through dietary manipulations, a disease analogous to human scurvy could be generated in guinea pigs. (Like

humans and unlike most animals, guinea pigs do not manufacture their own vitamin C.) When Holst and Frölich fed hay and oats (foods deficient in vitamin C) to guinea pigs, the animals developed scurvy, but when they were fed fresh fruits and vegetables, they remained healthy. In this way, Holst and Frölich were able to measure a food's ability to prevent scurvy.

While other scientists were trying to isolate vitamin C directly, Szent-Györgyi actually found the substance in the course of searching for something else. In the 1920's, his research centered on biological oxidation, that is, on how cells oxidize various food-

Albert Szent-Györgyi. (The Nobel Foundation)

stuffs. He was particularly entranced by the observation that some plants (apples and potatoes) turn brown after being cut and exposed to air, whereas others (oranges and lemons) experience no color change.

Szent-Györgyi suspected that a certain substance was controlling these color-change reactions, and he looked for it not only in fruits and vegetables but also in the adrenal cortex of mammals. He believed that the color change to a bronzelike skin in patients with Addison's disease (a disorder of the adrenal gland) was associated somehow with the color changes in plants. He hoped to isolate this substance, a powerful reducing agent, from the adrenal glands.

Unfortunately, his research was plagued with problems until he met the English biochemist Frederick Gowland Hopkins at a conference in Sweden in 1926. Hopkins was interested in vitamins and biological oxidation, and he invited Szent-Györgyi to the University of Cambridge to continue his research. Using many glands from oxen, Szent-Györgyi was able to separate a reducing agent from all other substances present. He also was able to obtain the same substance from orange juice and cabbage extracts, a result that his colleagues found most surprising.

"Godnose"

Through chemical analysis, he determined that the substance contained six carbon atoms and eight hydrogen atoms and that it was a carbohydrate related to the sugars. He initially wanted to name the substance "Ignose" (from the Latin *ignosco*, meaning "I don't know," and *ose*, the designating suffix for sugars). The editor of the *Biochemical Journal* thought that the name was too flippant, however, whereupon Szent-Györgyi suggested "Godnose," which was similarly rejected. Because the substance contained six carbon atoms and was acidic, he and his editor agreed on the name "hexuronic acid." News of the discovery of this acid was published in 1928.

At the time, scientists recognized five distinct vitamins. They had failed, however, to isolate any of them successfully. For this reason, it was not clear that Szent-Györgyi's hexuronic acid and vitamin C are the same substance.

In the fall of 1931, Joseph L. Svirbely, a postdoctoral student, arrived at Szeged, Hungary, where Szent-Györgyi had gone to continue his studies of vitamin C. Svirbely had done his doctoral studies on vitamin C under Charles Glen King at the University of Pennsylvania. King was trying, with limited success, to isolate vitamin C from lemon juice. He was testing his results with time-consuming experiments using animals.

Svirbely provided a bridge between King's work and Szent-Györgyi's. Szent-Györgyi had not previously tried to prove that hexuronic acid was identical to vitamin C because he did not enjoy working with animals. Furthermore, he was against vitamin research (he once said that vitamins were problems for the chef, not the scientist).

Nevertheless, when Svirbely mentioned that he could tell if something contained vitamin C or not, Szent-Györgyi gave him some of his hexuronic acid for experimentation. In a fifty-six-day test using guinea pigs, Svirbely established, in the fall of 1931, that the animals without hexuronic acid in their diets died with symptoms of scurvy, while the animals receiving hexuronic acid were healthy and free from scurvy. Further experiments in 1931-1932 proved once and for all that hexuronic acid and vitamin C are identical.

IMPACT

The isolation of vitamin C generated widespread comment and convinced most scientists that the long-sought vitamin had been found. Vitamin C's impact was deepened and extended by Szent-Györgyi's discovery in 1933 that Hungarian red peppers contained large amounts of the vitamin. Whereas previously biochemists could make only minuscule amounts of the material with great difficulty, Szent-Györgyi now could produce the substance in great quantities. In his lectures about his work, he liked to hold up a bottle containing several kilograms of the vitamin. To scientists accustomed to thinking of vitamins solely in extremely minute amounts, this was a surprising and enlightening experience.

In the 1930's, the League of Nations set up a committee to establish international standards for the vitamin, and the committee recommended that individuals ingest at least 30 milligrams each day to prevent scurvy. The vitamin came to be known as "ascorbic acid" for its property of combating scurvy. In the period during and after World War II (1939-1945), some scientists suggested that dosages larger than the recommended 30 milligrams would help keep humans in the best possible health. Many people, convinced that modern food processing was destroying vitamins, began to supplement their diets with vitamin pills, and some industries began to fortify their products with vitamins.

Beginning in 1965, Nobel-Prize-winning chemist Linus Pauling became interested in the megavitamin theory popularized by industrial chemist Irwin Stone in 1960. Pauling suggested that many maladies, from schizophrenia to cancer to the common cold, could be treated and prevented by large doses of vitamins. Pauling's books and articles created an ongoing controversy, guaranteeing that this fascinating substance, discovered through the efforts of Szent-Györgyi and others, will continue to provide subjects for rewarding scientific research well into the future.

See also Vitamin D.

FURTHER READING

Carpenter, Kenneth J. *The History of Scurvy and Vitamin C*. Cambridge, England: Cambridge University Press, 1986.
Friedrich, Wilhelm. *Vitamins*. New York: Walter de Gruyter, 1988.
Goldblith, Samuel A., and Maynard A. Joslyn, eds. *Milestones in Nutrition*. Westport, Conn.: AVI, 1964.
Leicester, Henry M. *Development of Biochemical Concepts from Ancient to Modern Times*. Cambridge, Mass.: Harvard University Press, 1974.
Moss, Ralph W. *Free Radical: Albert Szent-Györgyi and the Battle over Vitamin C*. New York: Paragon House, 1988.
Pauling, Linus. *Vitamin C, the Common Cold, and the Flu*. San Francisco: W. H. Freeman, 1976.
Waugh, William A. "Unlocking Another Door to Nature's Secrets—Vitamin C." *Journal of Chemical Education* 11 (February, 1934): 69-72.
—*Robert J. Paradowski*

VITAMIN D

THE SCIENCE: Elmer McCollum and collaborators established the existence of vitamin D, named it, and contributed to its use in the eradication of rickets.

THE SCIENTISTS:
Elmer Verner McCollum (1879-1967), American biochemist and nutritionist who carried out pioneering research on vitamin D, vitamin A, and the B vitamins
Thomas Burr Osborne (1859-1929), nutritionist under whom McCollum worked at the Connecticut Agricultural Station
John Howland (1873-1926), the physician-in-chief of pediatrics at The Johns Hopkins Hospital who collaborated with McCollum in several studies of rickets

RICKETS

Rickets (or rachitis) is a disease that causes abnormal bone formation, particularly in the long bones and the ribs. First described in the second century C.E. by Galen of Pergamum and Soranus of Ephesus, rickets was a widespread health problem until discovery and dissemination of the antirachitic factor, vitamin D. Elmer Verner McCollum and co-workers pioneered this effort in 1922, showing that the antirachitic factor was a distinctive substance. They named this substance vitamin D because it was the fourth vitamin to be discovered. The occurrence of rachitis is now rare—except in underdeveloped countries—as a result of the vitamin D fortification of food (especially milk) in the industrialized nations of the world.

THE EFFECT OF RICKETS

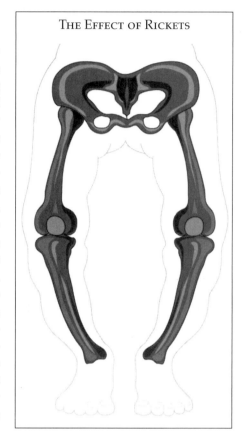

Rickets, which usually begins before age three, is caused by the improper and incomplete uptake of calcium into the fast-growing bones of children. The resultant insufficient calcification of these bones prevents them from hardening properly. Therefore, the bones of a rachitic child are so soft that they bend and twist into abnormal shapes. Furthermore, they will fracture easily. Fortunately, as afflicted children grow up, their bones harden, but the abnormal shapes are retained. Rickets is rarely fatal, but it produces several cosmetically unappealing conditions including curvature of the spine, bow legs, knock-knee, and chicken breast. Rickets sufferers are also unusually susceptible to the common cold, to bronchitis, and to pneumonia.

DAILY REQUIREMENTS

As may be expected, vitamin D is utilized in preventive chemotherapy, not in the correction of rickets. The two most common forms of the vitamin used in humans are calciferol (vitamin D_2) and cholecalciferol (vitamin D_3). These fatlike substances are derived from the steroids ergosterol and 7-dehydrocholesterol (7-DC), respectively. The human body converts 7-DC to cholecalciferol at the surface of the skin in a process that is energized by ultraviolet light from the Sun. Exposure of adults to normal amounts of sunlight causes enough vitamin D_3 production in the skin to make it unnecessary to add any vitamin D to their diet.

Children, however, require about 0.02 milligram of vitamin D per day in the diet if rickets is to be avoided. One way to administer the vitamin is

RICKETS ON THE RISE?

Rickets became rare in the United States after vitamin D supplementation of milk began in the mid-twentieth century. However, rates of incidence began to climb by the end of the century. Case studies of infants and toddlers with rickets found that all were exclusively breast-fed for at least the first six months of life and that few, if any, had received vitamin D supplementation.

It has been recognized for some time that the vitamin D content of human milk from healthy lactating women is low, approximately 22 International Units per liter. The 1998 edition of the *Pediatric Nutrition Handbook* of the American Academy of Pediatrics (AAP) recommended 400 International Units of vitamin D per day for breast-fed infants but indicated that only breast-fed babies with deeply pigmented skin needed vitamin D supplementation. Studies have shown that lactating women, regardless of skin color, who are deficient in vitamin D themselves produce milk with even lower concentrations of vitamin D.

In October, 2001, a meeting in Atlanta was sponsored by the Centers for Disease Control and Prevention (CDC) to examine the issue. Although many attendees of this meeting believed that universal vitamin D supplementation for all breastfed babies might be a good idea, others took the view that supplementation should be targeted toward specific groups. Some attendees worried that formula manufacturers would use supplement recommendations as a tool to encourage women to abandon breast-feeding. However, it was also made clear that in Canada, where vitamin D supplementation of all breast-fed children is advocated regardless of skin pigmentation, the number of women electing to breast-feed had increased.

By 2003, the American Cancer Society, the CDC, and the AAP had united in a campaign to cut the risk of skin cancer by cutting exposure to direct sunlight. The AAP now maintains that the recommended adequate intake of vitamin D cannot be met with human milk as the sole source of vitamin D for the breast-feeding infant and that some form of supplementation is needed.

as the cholecalciferol in cod-liver oil, a rich natural source of the vitamin. Fortification of milk with vitamin D_2 is more widespread today. It is important to note, however, that excess dietary vitamin D should be avoided. The Food and Nutrition Board of the Institute of Medicine has set the tolerable upper intake level for vitamin D at 25 micrograms (1,000 international units) for those up to 12 months old and 50 micrograms (2,000 international units) for children, adults, and pregnant or lactating women.

EARLY VITAMIN STUDIES

McCollum, who first identified vitamin D and named it, carried out many of the early important studies on this vitamin. McCollum's interest in biochemistry and vitamins began when he worked at the Connecticut Agricultural Experiment Station under Thomas Burr Osborne. This employment occurred during McCollum's doctoral training in organic chemistry at Yale University, and ended when he was awarded the Ph.D. in 1906. In 1907, McCollum was employed by the Wisconsin College of Agriculture, where he was assigned to investigate the chemical makeup of the food and excrement of dairy cattle. There he developed the first white rat colony in the United States devoted to use in the study of nutrition. Utilization of rats as experimental subjects allowed McCollum and his coworkers the opportunity to circumvent the complicated and tedious methodology that was required to study cattle and other large animals. This revolutionary concept of nutritional research was so successful that other scientists all over the world soon began to emulate McCollum's efforts. In six years, McCollum passed through the academic ranks from instructor to full professor.

In 1913, McCollum reported that rats fed "certain fat-deficient" diets exhibited a growth retardation that was reversed by feeding rats with "either extract of egg or of butter." By 1915, McCollum's

Elmer Verner McCollum. (National Library of Medicine)

research group had demonstrated that several trace substances, which McCollum called vitamins A and B, were necessary for normal health and growth in rats. Thus, McCollum helped to initiate the alphabetical names used in vitamin nomenclature. In 1917, McCollum became the chair of the department of chemistry and professor of biochemistry at the School of Hygiene and Public Health of The Johns Hopkins University in Baltimore. He continued his efforts to understand the vitamins and pioneered the study of vitamin D, for which he is best known, again using rats for his experiments.

THE LINE TEST

McCollum's pioneering identification of the existence of vitamin D in 1922 was accompanied by development of the line test for its measurement in foods. The line test begins with removal of bone sections (pieces of bone) from rats fed either normal, vitamin D-deficient, or vitamin D-supplemented diets. These bone sections are soaked in dilute solutions of light-sensitive silver nitrate. This treatment causes a silver compound to become a bone component wherever recent bone calcification has occurred. Exposure to light converts the silver compound to black, metallic silver in a process similar to that seen in photography. With normal bone, a very distinct black line is produced at the bone ends. No such line is seen is severe rickets, and indistinct lines are observed in healing cases of the disease. The test is "expressed with a scale of one to four, using plus and minus signs." It is viewed as both sensitive and accurate.

IMPACT

In 1922, rickets was a worldwide disorder that affected many children. Today, it has essentially been eradicated in developed nations despite fluctuations in its incidence. The successful treatment of the disease began when McCollum and coworkers produced evidence in 1922 that cod-liver oil contained a specific antirachitic chemical (vitamin D). As McCollum stated in *From Kansas Farm Boy to Scientist* (1964): "The demonstration of the existence of a vitamin which exerts a profound influence in directing the growth of bones proved to be of great public-health value."

McCollum demonstrated that this research stimulated great interest among many investigators. Furthermore, the discovery, coupled with the participation of prominent pediatricians in the effort, such as John Howland, led to rapid general acceptance by physicians of the efficacy of using cod-liver oil to prevent rickets. From that time on, the medical profession passed from haphazard use of the oil—in a skeptical fashion—to its routine use. As a result, rickets soon became rare.

Actualization of the existence of the antirachitic substance quickly led to isolation and characterization of vitamins D_2 and D_3. Subsequently, in the hands of other researchers, study of the pure vitamin began to show promise. First, it became possible to add vitamin D_2 to milk to ensure almost universal dissemination of the vitamin among the population of the industrialized countries. Next, it was shown that vitamin D_2 (or D_3) functioned after conversion as another chemical that was actually a hormone (hormone D).

The form of hormone D made by the body, from vitamin D_3 is called 1,25-dihydroxycholecalciferol. Hormone D acts by stimulating rapid intestinal reabsorption of calcium via a protein. This calcium resorption minimizes calcium loss in the feces and prevents the bone decalcification that results in rickets.

Additional examination of the action of vitamin D has led to better understanding of the processes of bone deposition and resorption as well as to explanation of the interrelationships between hormone D and other calcium-controlling substances (such as calcitonin and parathyroid hormone) made by the body. Such investigations have also led to the realization that bone is not simply a "dead," body-support matrix. Rather, bone is a vital, live tissue that can produce dissolved calcium in the blood to serve many purposes.

This realization has had further ramifications, and it is clear that calcium serves as a biological signal in life processes that include control of the blood pressure, blood clotting, nerve impulse transmission, and muscle contraction. Therefore, the acorn of McCollum's efforts had produced a mighty oak tree of intertwined information about life. This information now promises eventual answers to many elusive but fundamental problems of life science that are clearly associated with calcium.

See also Vitamin C.

FURTHER READING

Funk, Casimir. *The Vitamins*. Baltimore: Williams & Wilkins, 1922.
McCollum, Elmer V. *From Kansas Farm Boy to Scientist*. Lawrence: University of Kansas Press, 1964.
_____. *History of Nutrition*. Boston: Houghton Mifflin, 1957.
_____. *The Newer Knowledge of Nutrition*. 2d ed. New York: Macmillan, 1922.

Pike, J. Wesley, Francis H. Glorieux, and David Feldman, eds. *Vitamin D.* 2d ed. Boston: Academic Press, 2004.

—*Sanford S. Singer*

Voyager Missions

THE SCIENCE: The Voyager probes executed the first Grand Tour in planetary exploration by successively encountering Jupiter, Saturn, Uranus, and Neptune. Such a tour, using the "planetary-gravity-assist" technique to travel from planet to planet, is possible only once every 175 years.

THE SCIENTISTS:

Gary A. Flandro, discoverer of Grand Tour alignments of the outer planets

Charles E. Kohlhase, Principal Mission designer

Donald M. Gray (b. 1929), the navigation team chief of Voyager, NASA, Jet Propulsion Laboratory

Harris M. Schurmeier, Voyager project manager through development phase

John R. Casani (b. 1932), Voyager project manager from launch to Jupiter encounters

Raymond L. Heacock (b. 1928), Voyager project manager for Jupiter and Saturn encounters

Richard P. Laeser, Voyager project manager for Uranus encounter

Norman Ray Haynes (b. 1936), Voyager project manager for Neptune encounter

Edward C. Stone, Jr., Voyager project scientist

Ellis D. Miner (b. 1937), assistant project scientist

Andrei B. Sergeyevsky, principal trajectory designer for the Neptune encounter

Bradford A. Smith, principal investigator, imaging science experiment

G. Leonard Tyler, principal investigator, radio science experiment

Laurence Soderblom (b. 1944), expert on Galilean satellites

THE GRAND TOUR

Often referred to as the "grand tour," the flights of Voyagers 1 and 2 passed the large outer planets of the solar system and returned detailed

Artist's rendition of the Voyager spacecraft. (NASA)

and valuable scientific information. The mission spanned twelve years, from 1977 to 1989.

As the outer planets orbit the Sun, they are, once in a great while, aligned in a pattern that presents an opportunity for a spacecraft launched from Earth to fly past each one of them. Such an alignment occurs only once every 175 years. The fact that the last such favorable planet alignment had occurred during the term of U.S. president Thomas Jefferson was cited by those in favor of the Voyager program. The Voyager missions required a spacecraft designed not only to survive intense radiation but also to be able to detect and react to any problems that might arise. Earth commands would require far too much time for corrective action because of the much greater distance to the outer planets.

Voyager 1

Voyager 1 was launched on September 5, 1977. It made its closest approach to Jupiter on March 5, 1979, a year and a half after launch and four months ahead of Voyager 2. (Voyager 2 had been launched on August 20, 1977, sixteen days ahead of Voyager 1.) Passing the planet, it encountered the moons Amalthea, Io, Ganymede, and Callisto and passed to within one million kilometers of Europa, another of Jupiter's moons.

Voyager 1 encountered Saturn and its moons in early November, 1980. The closest approach to the outer moon, Titan, occurred on November 11. It swung beneath Saturn's "ring" system and behind the planet, as viewed from Earth, making its closest approach to the moon Iapetus on November 14, 1980. It then began to travel deeper into space in the direction of the constellation Ophiuchus.

VOYAGER 2

Voyager 2 arrived at Jupiter in July, 1979, with a different trajectory from that of Voyager 1 and at different angles, which permitted photographs of the opposite hemispheres of Callisto and Ganymede, high-resolution images of Europa, and shots of Jupiter's ring. Just as the eighty-hour gravitational tug of Jupiter had propelled the tiny Voyager 2 toward Saturn, so now giant Saturn pulled it into a new direction: directly toward Uranus, a journey of almost four and a half years. As this outward journey from Saturn began, Voyager 2 now became the true trailblazer, going where no spacecraft had ever gone before. It arrived at Uranus on schedule. Then the gravitational pull of that planet bent Voyager's trajectory toward its last assignment, planet Neptune.

Voyager 2 reached Neptune on August 24, 1989, having traveled about seven billion kilometers. That distance was so great that the probe's radio signals required about four hours just to reach Earth. The flyby of Neptune and its moons was a complete success. Approaching the planet, Voyager 2 swung over its north pole and then encountered Triton, Neptune's moon, which has an atmosphere. It then began its one-way trip into deep space, transmitting data as long as its systems remained active.

IMPACT

Both the technological and the scientific successes of the Voyager missions were remarkable and can hardly be overstated. The mechanical and electronic components of the spacecraft functioned exceedingly well over the twelve-year span (1977-1989).

Voyager 1 created an explosion of excitement when it detected a ring around Jupiter. The discovery was unexpected, since no evidence had ever been presented supporting its possible existence. Mission planners had agreed to devote a single photograph to a "one-shot" search for a ring, and luckily that was enough. As for Jupiter's atmosphere, Voyager 1 sent back high-resolution images of the Great Red Spot, which has been observed in Jupiter's upper atmosphere since the early 1800's, showing exquisite detail.

Callisto, the second largest of Jupiter's moons and the most heavily cratered, was determined to have a diameter about one and one-half times that of Earth, although its density is only about one-third as great. This suggested that Callisto is about one-half water and ice. The fact that no deep craters were found supports this model, since water-ice walls are not strong enough to stand very high and tend to flow down into the crater floor.

Though scientists expected to see craters on Io, none were seen. The extremely active volcanoes on Io were the most startling discovery of the mission, and Io is now believed to be the most geologically active object in the solar system. In appearance, Europa exhibits extensive cracks and faults quite unlike any seen before. They are very long, crisscrossing like blood vessels. Triton, orbiting Neptune, showed complex surface features resembling a cantaloupe, signs of past volcanic activity, and a large polar ice cap.

As of 2005, the two Voyager spacecraft were continuing their scientific exploration of interstellar space. Voyager 1 was 13.5 billion kilometers from Earth, nearly three decades after launch, and was traveling away from the solar system at 21 kilometers per second. Voyager 2 was 11 billion kilometers from Earth, traveling at 29 kilometers per second.

See also Cassini-Huygens Mission; Earth Orbit; Galileo Mission; International Space Station; Jupiter's Great Red Spot; Kepler's Laws of Planetary Motion; Mars Exploration Rovers; Moon Landing; Oort Cloud; Planetary Formation; Pluto; Saturn's Rings; Solar Wind; Space Shuttle; Wilkinson Microwave Anisotropy Probe.

FURTHER READING

Cooper, Henry S. F., Jr. *Imaging Saturn: The Voyager Flights to Saturn.* New York: H. Holt, 1985.

Davis, Joel. *Flyby: The Interplanetary Odyssey of Voyager 2.* New York: Atheneum, 1987.

Harland, David M. *Jupiter Odyssey: The Story of NASA's Galileo Mission.* London: Springer-Praxis, 2000.

_____. *Mission to Saturn: Cassini and the Huygens Probe.* London: Springer-Praxis, 2002.

Hartmann, William K. *Moons and Planets.* 5th ed. Belmont, Calif.: Thomson Brooks/Cole, 2005.

Irwin, Patrick G. J. *Giant Planets of Our Solar System: Atmospheres, Composition, and Structure.* London: Springer-Praxis, 2003.

Morrison, David. *Voyages to Saturn.* NASA SP-451. Washington, D.C.: National Aeronautics and Space Administration, 1982.

Morrison, David, and Jane Samz. *Voyage to Jupiter.* NASA SP-439. Washington, D.C.: Government Printing Office, 1980.

Poynter, Margaret, and Arthur L. Lane. *Voyager: The Story of a Space Mission.* New York: Macmillan, 1981.

—*Richard C. Jones and William J. Kosmann*

WATER

THE SCIENCE: After discovering "inflammable air" (hydrogen), Henry Cavendish investigated its properties, eventually finding that the product formed when it burned in "dephlogisticated air" (oxygen) was pure water.

THE SCIENTISTS:

Henry Cavendish (1731-1810), English natural philosopher best known for his research on gases and the nature of water

Joseph Priestley (1733-1804), English scientist and Unitarian minister who discovered oxygen and several other new gases

James Watt (1736-1819), English inventor of an improved steam engine who gave an interpretation of water's nature

Antoine-Laurent Lavoisier (1743-1794), French chemist who interpreted water as a compound of hydrogen and oxygen

Sir Charles Blagden (1748-1820), Cavendish's assistant during the years that he studied hydrogen and the nature of water

FROM GAS TO WATER

In the seventeenth and eighteenth centuries, scientists such as Robert Boyle noticed that a flammable gas was generated when acids were added to metals. The Englishman Henry Cavendish was sufficiently intrigued by this gas to study it comprehensively. He prepared it with various metals (iron, zinc, and tin) and acids (what is now known as hydrochloric and sulfuric acids). Using two different methods, he determined the gas's specific gravity, finding it was nearly nine thousand times lighter than water and about one-fourteenth the weight of common air. When he introduced a flame into a mixture of this gas and ordinary air, the gas burned bright blue, and so he called it "inflammable air from the metals," which was later shortened to "inflammable air"; its modern name is hydrogen.

Because Cavendish, like many scientists of his time, believed in the phlogiston theory—which posited that every combustible material contained phlogiston—and because inflammable air burned with no residue, Cavendish interpreted this new gas as phlogiston. In 1766 he published his findings in a tripartite paper in which each part dealt with a specific gas prepared by a certain process: (1) inflammable air from metals and acids, (2) fixed air (carbon dioxide) from alkalis and acids, and (3) "mixed airs" from organic materials by fermentation or putrefaction.

Cavendish's report on inflammable air stimulated Joseph Priestley, who, in 1781, put an electric spark through a mixture of inflammable air and common air in a dry glass container and noticed that the inside of the glass container became coated with moisture. Neither Priestley nor a colleague who helped him understood what they had done, but Cavendish, who repeated their experiment in a systematic and quantitative way, did. During the summer of 1781 he found that all the inflammable air and about one-fifth of the ordinary air had ceased being gases in forming what he discovered was pure water.

Cavendish and Priestley routinely interacted, and so Cavendish was aware of a new gas, "dephlogisticated air" (oxygen), that Priestley had discovered in 1774. He was therefore curious about what would happen when he sparked various mixtures of inflammable air and dephlogisticated air. After several trials he established that a two-to-one ratio of inflammable to dephlogisticated air led to the complete conversion of these gases to water. Although he was the first scientist to establish this experimental fact, his interpretation of his results was confusing. The obvious explanation was to see water as the union of these two gases, but Cavendish was a phlogistonist and still tied, in a way, to the old idea of water as a chemical element on its own (we now know it to be the compound of two hydrogen atoms joined to one atom of oxygen, or H_2O). For Cavendish, inflammable air was either phlogiston or "phlogisticated water" (water united to phlogiston). Dephlogisticated air, on the other hand, was water deprived of its phlogiston. Therefore Cavendish saw water as preexisting in the combining gases, and the spark-induced reaction simply revealed what had previously been hidden.

LAVOISIER'S NEW CHEMISTRY

Even though Cavendish did not publish his experimental results and interpretation until 1784, scientists in England and France learned about them. For example, in 1783 Sir Charles Blagden, Cavendish's assistant, made a trip to Paris during which he met Antoine-Laurent Lavoisier and

LAVOISIER, WATER, AND CHEMICAL ELEMENTS

Until the late eighteenth century, water was considered to be a chemical element rather than a compound. This 1783 report prepared for the Royal Academy relates how Antoine-Laurent Lavoisier advanced Henry Cavendish's experiments to identify water a compound of hydrogen and oxygen:

M. Cavendish . . . observed that if one operates in dry vessels a discernible quantity of moisture is deposited on the inner walls. Since the verification of this fact was of great significance to chemical theory, M. Lavoisier and M. [Pierre-Simon] de la Place proposed to confirm it in a large-scale experiment. . . . The quantity of inflammable air burned in this experiment was about thirty pints [pintes] and that of dephlogisticated air from fifteen to eighteen.

As soon as the two airs had been lit, the wall of the vessel in which the combustion took place visibly darkened and became covered by a large number of droplets of water. Little by little the drops grew in volume. Many coalesced together and collected in the bottom of the apparatus, where they formed a layer on the surface of the mercury.

After the experiment, nearly all the water was collected by means of a funnel, and its weight was found to be about 5 gros, which corresponded fairly closely to the weight of the two airs combined. This water was as pure as distilled water.

A short time later, M. Monge addressed to the Academy the result of a similar combustion . . . which was perhaps more accurate.

Antoine-Laurent Lavoisier

(Library of Congress)

He determined with great care the weight of the two airs, and he likewise found that in burning large quantities of inflammable air and dephlogisticated air one obtains very pure water and that its weight very nearly approximates the weight of the two airs used. Finally . . . M. Cavendish recently repeated the same experiment by different means and that when the quantity of the two airs had been well proportioned, he consistently obtained the same result.

It is difficult to refuse to recognize that in this experiment, water is made artificially and from scratch, and consequently that the constituent parts of this fluid are inflammable air and dephlogisticated air, less the portion of fire which is released during the combustion.

Source: "Report of a Memoir Read by M. Lavoisier at the Public Session of the Royal Academy of Sciences of November 12, on the Nature of Water and on Experiments Which Appear to Prove That This Substance Is Not Strictly Speaking an Element but That It Is Susceptible of Decomposition and Recomposition." *Observations sur la Physique* 23 (1783): 452-455. Translated by Carmen Giunta.

informed him about how Cavendish and he had made pure water from two new gases. Lavoisier quickly realized the implications of their results for his new theory of chemistry, which was overturning accepted paradigms of elemental matter: Lavoisier had been conducting investigations into the four "elements"—earth, air, fire, and water—and by 1779 he was disproving the phlogiston theory and in the process identifying a list of more than thirty chemical elements. For this he would become known to posterity as the father of modern chemistry.

In November of 1783, Lavoisier reported to the French Academy of Sciences on experiments that he and Pierre-Simon Laplace had performed demonstrating that water was not an element but a compound of hydrogen and oxygen. Lavoisier failed to mention the stimulus he had received from the research of Cavendish, who did not publish his results until 1784. In this later publication Cavendish was able to complete his earlier studies by showing that the gas that was left behind when dephlogisticated air was removed form common air was a colorless gas in which mice died and a candle would not burn; this gas was what Lavoisier called "azote" and others called "nitrogen."

THE WATER CONTROVERSY

Because the compound nature of water was such a significant discovery and because so many people contributed in one way or another to it, a "water controversy" developed. The debate was basically a priority dispute. Because both Priestley, who could have made a claim but never did, and Cavendish, whose introverted personality ill suited him to controversy, stayed on the sidelines, the contending parties in the first phase of the water controversy were James Watt and Antoine Lavoisier. Watt became involved because Priestley told him about his dew-forming experiments and Watt then circulated his interpretation of Priestley's results to Royal Society members.

When Watt learned of Cavendish's and Lavoisier's reports on water's nature, he accused Cavendish of plagiarizing his ideas and Lavoisier of plagiarizing Cavendish's experiments. For his part, Cavendish was willing to give credit to Lavoisier for interpreting the composition of water in terms of the oxygen theory. Although most historians of science appreciate Lavoisier's contributions, they criticize him for neglecting to give credit to Cavendish. These scholars also find Watt's claims confused and his interpretation derivative. Indeed, they bestow on Cavendish, the least contentious of the claimants, the lion's share of the honor for finding water's true nature.

IMPACT

Some scholars see Cavendish as Britain's preeminent eighteenth century scientist, between Isaac Newton in the seventeenth century and James Clerk Maxwell in the nineteenth. Cavendish's studies of what he called "factitious airs" (those contained in solids) were models of a rigorously quantitative approach to chemistry. Future chemists would use his methods for generating, collecting, transferring, and measuring gases and for determining their unique characteristics. Using these methods he contributed significantly not only to discovering the composition of water but also to clarifying the nature of such compounds as nitric acid. His quantitative studies of the specific combining volumes of the gases necessary to form water constituted an important step toward the law enunciated by Joseph Louis Gay-Lussac in 1809 that the ratios of the volumes of reacting gases are always small whole numbers.

Cavendish's experimental contributions were much more important than his theoretical contributions, and his adherence to the phlogiston theory hampered his understanding of his experimental results almost to the end of his life, when he finally began to see some value in the new chemistry of Lavoisier.

The water controversy was significant because of what it revealed about the changing nature of science. Before the eighteenth century scientists tended to work alone, and their discoveries were often seen as a consequence of their individual genius. In the eighteenth century scientific discoveries increasingly involved many talented individuals working in concert with or in cognizance of many others. Inevitably more than one scientist would sometimes arrive at the same conclusion or discovery at about the same time. Some scholars attribute the water controversy to the casual way in which scientific data were then gathered, dated, and reported. Other scholars point to nationalism as a factor in the water controversy, especially as it continued in the nineteenth century after the deaths of the original contenders. French and British scholars, using newly available primary sources, argued about the credit that should be given to Watt and Cavendish.

One significant by-product of the study of Cavendish's papers was the role that some his data played in the discovery of a new element in 1894. When Cavendish in the eighteenth century had removed oxygen and nitrogen from ordinary air, he found a small bubble of gas still remaining. In the late nineteenth century this bubble of gas was shown to be argon, a new noble gas, a belated testimony to the meticulousness of Cavendish's experimental prowess.

See also Atomic Theory of Matter; Carbon Dioxide; Definite Proportions Law; Oxygen; Photosynthesis.

Further Reading

Holmes, Frederic Lawrence. *Antoine Lavoisier, the Next Crucial Year: Or, The Sources of His Quantitative Method in Chemistry.* Princeton, N.J.: Princeton University Press, 1998.
Jaffe, Bernard. *Crucibles: The Story of Chemistry.* New York: Dover, 1998.
Jungnickel, Christa, and Russell McCormmach. *Cavendish: The Experimental Life.* Lewisburg, Pa.: Bucknell, 1999.
Miller, David Philip. *Discovering Water: James Watt, Henry Cavendish, and the Nineteenth Century "Water Controversy."* Burlington, Vt.: Ashgate, 2004.
Strathern, Paul. *Mendeleyev's Dream: The Quest for the Elements.* New York: Berkley Books, 2000.

—*Robert J. Paradowski*

WAVE-PARTICLE DUALITY OF LIGHT

THE SCIENCE: Louis de Broglie provided a mechanical explanation for the wave-particle duality of light.

THE SCIENTISTS:
Louis de Broglie (1892-1987), French prince, historian, and physicist who won the 1929 Nobel Prize in Physics
Niels Bohr (1885-1962), Danish physicist who won the 1922 Nobel Prize in Physics
Erwin Schrödinger (1887-1961), Austrian physicist who won the 1933 Nobel Prize in Physics

BOTH A WAVE AND A PARTICLE

In the early 1900's, scientists were having difficulty describing the nature of light. For a long time, light had been regarded as acting like a particle. In the late nineteenth century, the wavelike nature of light had been demonstrated. Early in the twentieth century, however, this belief was shifted again by experiments that confirmed the particle nature of light. The wave-particle duality of light was an experimental phenomenon in search of a theory.

At the beginning of the twentieth century, German physicist Max Planck had used the concept of the wave nature of light to explain blackbody radiation (radiation from a theoretical celestial body capable of completely absorbing all radiation falling on it). As a wave, light has a wavelength (the dis-

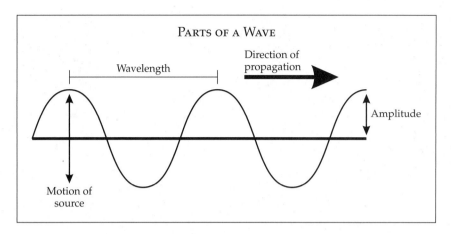

tance between crests) and a corresponding frequency (the number of crests passing a point in a given amount of time). Planck had shown that light of a particular frequency had a definite amount of energy; that is, energy is quantized. This seemed to favor the belief that light was wavelike in nature.

Nevertheless, five years later, in 1905, American physicist Albert Einstein reasoned that light behaved like particles. Einstein used Planck's theory of quantized light to explain why light striking the surface of certain metals resulted in the ejection of electrons from that metal (the photoelectric effect), but only when this involved certain frequencies of light. He pictured the light striking the metal surface as particles of light, or photons, with sufficient energy to knock off electrons.

The wave-particle nature of light was constantly debated and seemed dependent upon the experiment being performed. For example, the dispersion of white light into its component colors by a prism is a result of the wave nature of light. By contrast, the ability of a stream of photons to eject electrons from a metal surface points to the particle nature of light. Einstein had shown by his relativity theory that light could behave like both waves and particles and that the physical properties of each nature were related. He showed that the momentum of the photon (a particle property) was related to the wavelength of the light (a wave property). Einstein's results demonstrated that light has wave and particle duality.

A SNAKE SWALLOWING ITS OWN TAIL

Louis de Broglie had been studying Planck's theories of quantized light and Einstein's wave-particle concept of light. He wrote several papers calling attention to the dual behavior of light. De Broglie wished to provide a mechanical explanation for the wave-particle duality. Thus, he needed to find a mechanical reason for a particle—the photon—to have an energy that was determined by a wave, or rather by the frequency of that wave. While he was thinking about light, the idea occurred to de Broglie that matter (a particle) might have a wave nature also.

At about this time, Niels Bohr had revealed a theory for the electronic structure of atoms. Bohr's theory was that the electrons in an atom were restricted to particular energy levels and positions called "orbitals." Only by exact additions of unit amounts of energy could the energy and orbital of an electron be changed.

De Broglie was struck by the analogy of Bohr's orbital energies to standing waves. As a result, de Broglie discovered an example of wave-particle duality in matter.

De Broglie used his explanations of the wave-particle duality of matter in writing his doctoral dissertation in physics. He presented his dissertation before the Faculty of Sciences at the University of Paris in 1923. His theory demonstrated that matter, like light, had a wavelike nature.

De Broglie noticed that the momentum of the electron orbitals proposed by Bohr were whole number units of a fundamental quantity, Planck's constant. He knew that standing waves had unit changes in their momenta also. A standing wave can be thought of as a string, fixed at both ends, that is plucked. The string will oscillate back and forth, yet some points will remain at rest. The number of rest points will increase as the frequency of the vibration increases. De Broglie reasoned that Bohr's orbitals could therefore be seen as a circular string, a snake swallowing its own tail.

Moreover, de Broglie discovered that the matter waves he had proposed fit Bohr's electron orbits exactly. He also found that the momenta and wavelengths of his matter waves were related, like those of light. He had succeeded in explaining Bohr's orbits: Each orbit was a steady wave pattern, and these orbits had determined and fixed sizes so that these distinct "quantized" wave patterns could exist.

When de Broglie somewhat reluctantly submitted his dissertation, the faculty at the University of Paris was unsure of the use of strings to explain Bohr's orbits and asked Einstein to judge the acceptability of the dissertation. Einstein confirmed that it was sound. The thesis was accepted, and later de Broglie was awarded the Nobel Prize.

IMPACT

De Broglie's waves had offered a picture of what was occurring inside an atom. A way to visualize the shifting patterns of the wave was needed when the atom changed energy and produced light.

Erwin Schrödinger, an Austrian physicist, found a mathematical equation that explained the changing wave patterns inside an atom. Schrödinger's equation provides a continuous mathematical description of the wave-particle duality of matter. He viewed the atom as analogous to de Broglie's vibrating string. The movement of the electron from one orbit to another was a simple change in the frequency of the standing waves of the string. In a musical string, this occurs as the harmony of two wave patterns, the result being the differences in the frequency of the two waves.

The understanding of the wave-particle duality of matter, as modeled by Schrödinger's equation, was instrumental in the founding of quantum physics. Quantum physics has been responsible for many of the technological advances in the twentieth century. These advances are traceable to de Broglie's pronouncement of the wave-particle duality of matter.

See also Alpha Decay; Atomic Structure; Compton Effect; Diffraction; Electrons; Exclusion Principle; Grand Unified Theory; Heisenberg's Uncertainty Principle; Lasers; Optics; Photoelectric Effect; Quantized Hall Effect; Quantum Chromodynamics; Quantum Mechanics; Schrödinger's Wave Equation; Superconductivity; X-Ray Fluorescence.

Further Reading

Guillemin, Victor. *The Story of Quantum Mechanics*. New York: Charles Scribner's Sons, 1968.

Hoffmann, Banesh. *The Strange Story of the Quantum*. 2d ed. New York: Dover, 1959.

Jammer, Max. *The Philosophy of Quantum Mechanics*. New York: John Wiley & Sons, 1974.

McQuarrie, Donald A. *Quantum Chemistry*. Mill Valley, Calif.: University Science Books, 1983.

Wolf, Fred Alan. *Taking the Quantum Leap*. San Francisco: Harper & Row, 1981.

—*Scott A. Davis*

Weather Fronts

The Science: Vilhelm Bjerknes's model of the atmosphere emphasized the idea of "fronts," those boundaries along which masses of warm and cold air clash and converge to produce the weather.

THE SCIENTISTS:
Vilhelm Bjerknes (1862-1951), Norwegian meteorologist
Jacob Bjerknes (1897-1975), Norwegian meteorologist
Carl-Gustav Arvid Rossby (1898-1957), Swedish American
   meteorologist
Tor Bergeron (1891-1977), Swedish meteorologist

GATHERING INFORMATION

Vilhelm Bjerknes, a Norwegian geophysicist and meteorologist, knew in the early 1900's that accurate forecasting of weather required much information. He also knew that local weather was tied to the global circulation of the atmosphere. While teaching at Stockholm University from 1895 to 1907, he proposed that movements in the atmosphere are stimulated by heat from the Sun. At the same time, these movements radiate heat as air masses rub up against one another, causing friction.

Bjerknes was motivated by the need for improved weather prediction for commercial fishing and agriculture. In part, the urgent need for better domestic food production arose from restrictions of imports and communications as a result of World War I (1914-1918). He persuaded the Norwegian government to help set up strategically located observing stations. In addition to the stations, he founded a school at Bergen that attracted meteorologists from all over the world, including his son Jacob Bjerknes; Carl-Gustav Arvid Rossby, a Swedish American meteorologist; and Tor Bergeron, a Swedish meteorologist.

Weather types and changes, along with moving masses of air and their interaction, have been studied and noted for centuries. In the nineteenth century, Luke Howard, an English physicist, had written of northerly and southerly winds blowing alongside each other, with the colder wedging in under the warmer and the warmer gliding up over the colder and causing extensive and continued rains. In 1852, evidence had been found of a polar wind advancing under a warm, nearly saturated tropical wind and pushing it upward producing cumulus clouds.

EXPLAINING THE WEATHER

Vilhelm Bjerknes was a pioneer in the development of a mathematical theory of fronts and their effects. In addition, along with his son, he was the first to study extratropical cyclones and use them to forecast the weather. Extratropical cyclones are cyclones that may cross an ocean in ten days, lose most of their intensity, and then develop again into large and vigorous

One of the most terrifying weather fronts is that of a hurricane—here, Hurricane Fran in 1996. (NASA/GSFC)

storms. In the years following World War I, Norwegian meteorologists had a fairly good understanding of the action in the big storms sweeping across the Atlantic. From this knowledge, Vilhelm Bjerknes theorized that the main idea in storm development is a clashing of two air masses, one warm, the other cold, along a well-defined boundary, or front.

Aside from the idea of storm fronts, his view of cyclone development produced another important idea. At the beginning of the life cycle of a storm, there is an undisturbed state in which cold and warm air masses flow side by side, separated by a front. Each air mass flows along its side of the front until some of the warmer air begins to invade the cooler air, leading to a wave disturbance. This disturbance spreads and grows, creating low-pressure areas at the tip of the wave. Air motions try to spiral into these areas, and both fronts begin to advance. The cold air generally moves faster, catching up with and moving under the lighter warm air. As the storm grows deeper, the cold front becomes more pronounced. The whole process—from the time the polar air meets the northward-flowing warm air to the point at which the area of low pressure is filled completely—is known as the "life cycle of a frontal system." This description is based on the wave theory, which was originally developed by Vilhelm Bjerknes in 1921.

In 1919, when this work began, upper atmosphere studies were limited by the lack of knowledge of such things as radar images, lasers, computers, and satellites. Vilhelm Bjerknes showed that the atmosphere is composed of distinct masses of air meeting at various places to produce different meteorological effects. He published the study *On the Dynamics of the Circular Vortex with Applications to the Atmosphere and Atmospheric Vortex and Wave Motion* in 1921.

IMPACT

The weather forecasting stations established by Vilhelm Bjerknes in the 1920's were a monumental accomplishment, considering the limited amount of information and the lack of high-speed, worldwide communications. All the computations were done without the assistance of a computer or modern weather satellites to analyze and model the data. Today, these and other tools have made it possible to compile data into real-time images of the current weather patterns around the globe. It was pioneers such as Vilhelm Bjerknes who paved the way.

See also Atmospheric Circulation; Atmospheric Pressure; Chaotic Systems; Fractals; Ionosphere; Stratosphere and Troposphere.

FURTHER READING

Bates, Charles C., and John F. Fuller. *America's Weather Warriors, 1814-1985.* College Station: Texas A&M Press, 1986.

Bjerknes, Vilhelm, and Johann Sandstrøm. *Dynamic Meteorology and Hydrography.* Vol. 1. Washington, D.C.: Carnegie Institution, 1910.

Friedman, Robert Marc. *Appropriating the Weather: Vilhelm Bjerknes and the Construction of a Modern Meteorology.* Ithaca, N.Y.: Cornell University Press, 1989.

Holmboe, Jorgen, George E. Forsythe, and William Gustin. *Dynamic Meteorology.* New York: John Wiley, 1945.

Kutzbach, Gisela. *The Thermal Theory of Cyclones: A History of Meteorological Thought in the Nineteenth Century.* Boston: American Meteorological Society, 1979.

—*Earl G. Hoover*

WILKINSON MICROWAVE ANISOTROPY PROBE

THE SCIENCE: The Wilkinson Microwave Anisotropy Probe (WMAP), a space-based astronomical observatory designed to measure the "echo" or "afterglow" of the big bang known as cosmic microwave background, provided the first accurate measure of the age of the universe, changed the way astronomers think about the earliest star formation, and supported some of the leading cosmological theories.

THE SCIENTISTS:

David T. Wilkinson (1935-2002), project scientist, Princeton University
Robert H. Dicke (1916-1997), experimental physicist at Princeton University

AFTER THE BIG BANG

Astronomer George Gamow speculated in 1948 on the existence of an "echo" of the early events in the history of the universe. After the big bang, or moment of creation, light and matter were bound together in such a way that the universe was opaque to light. When the temperatures from the hot big bang cooled to approximately 300 Kelvins (degrees above absolute zero), matter and light separated in what is called "last scattering." The light from that moment is still reaching Earth today but has been red-shifted to 2.73 Kelvins given the large distance. This cosmic microwave background radiation (CMB) was accidentally discovered in 1965 by astronomers and Bell Telephone Laboratory scientists Arno A. Penzias and Robert W. Wilson. They discovered this seemingly isotropic radiation while studying emissions from the Milky Way. This discovery earned them the Nobel Prize in Physics in 1978.

Astronomers continued to study the CMB for decades. However, it puzzled them in that the measured temperature was perfectly smooth in all directions. After all, if the universe was perfectly smooth, then clumps of matter like galaxies, stars, and planets could not have formed. Hence, although the existence of the CMB supported the big bang theory, its uniformity did not fit in with observations of the modern universe. Finally, in 1992, the Cosmic Background Explorer satellite (COBE, which had been launched on November 9, 1989) measured anisotropies, or irregularities, in the CMB temperature on small scales. Although this fuzzy picture did not provide much scientific detail, the seeds of the early galaxies had finally been detected.

Mapping the Universe

At Princeton University, David Wilkinson had been working with Robert Dicke on a receiver to detect the CMB. Originally named the Microwave Anisotropy Probe (MAP) and later the Wilkinson Microwave Anisotropy Probe (WMAP), it was proposed to NASA in 1995 to follow up on the COBE discoveries and was authorized in 1997. It became a space-based observatory and the most successful observational cosmology project to date. Launched on June 30, 2001, this unique orbiting radio telescope began a journey that would answer many fundamental questions about the evolution of the universe.

WMAP featured two radio telescopes 140° apart in order to map the temperature of the sky in all directions. This design allowed for differential mapping—or subtracting the temperature in one area of the sky from the temperature at another point—which allows for subtraction of false signals. In other words, the relative temperatures of different regions are measured. In this way, WMAP can achieve a sensitivity of 0.000020 Kelvin. This is necessary in order to determine the tiny density variations in a radiation field that is only 2.73 Kelvins. The temperature of an object or energy field is related to the peak wavelength of the emission. Since the CMB temperature is so low, it has a low energy and therefore peaks at long wavelengths, specifically in the microwave region of the electromagnetic spectrum.

The cosmic background radiation is easily washed out by "foreground"

The Wilkinson Microwave Anisotropy Probe provided data that made possible this spectacular map of the early universe, including the very first anisotropies, which grew into the stars and galaxies that exist today. (NASA)

objects, mainly galaxies, gas clouds, or human-made signals. Therefore, WMAP operates at five frequencies: 22, 30, 40, 60, and 90 gigahertz. Because many ground-based telescopes operate at these frequencies, astronomers know much about objects in the radio sky and can subtract that radiation from the CMB radiation in the WMAP data with great precision.

The result was a spectacular map of the early universe and the very first anisotropies that grew into the stars and galaxies that exist today. The map of the universe that was produced from the first-year results has become a mainstay of science media. The map is color-coded to show tiny temperature variations. The warmer temperature spots indicate density clumps, and the cooler spots indicate empty space. These clumps vary in temperature by 0.0002 Kelvin. From these data, many conclusions could be made about the nature, history, and future of the universe.

How Old Is the Universe?

The data gleaned from the WMAP mission helped astronomers pin down the most important cosmological parameters. Using sophisticated modeling techniques, astronomers are now able to start with a clear early picture of the universe and test different evolutionary models until the result resembles the current universe. For the first time, the age of the universe has been pinpointed with incredible accuracy at 13.7 billion years, because the Hubble constant, or a measure of the expansion rate of the universe, has been determined to be 71 (km/sec)/Mpc. (A megaparsec, or Mpc, is approximately 3.26 million light-years; kilometers per second is denoted as km/sec.)

Also, the first stars seem to have turned on, or have begun nuclear fusion, 200 million years after the big bang—long before anyone had originally thought.

WMAP also confirmed that the geometry of the universe is flat. That is, the Euclidean geometry that is learned in high school applies over large scales. Other theories had surmised that the universe could be curved. In these strange geometries, parallel lines could eventually intersect or diverge over long distances.

Dark Matter, Dark Energy

Our universe has its own strange qualities, nevertheless. This flat geometry, along with other cosmological data to date, suggest that only 4 percent of the matter in the universe is the matter that makes up stars, planets, and humans, or baryonic matter, while 23 percent of the matter in the universe is known as dark matter—that which has not yet been detected but exerts gravitational forces on nearby objects.

An even stranger thing, known only as dark energy, forms 73 percent of the universe. Theorists do not yet know of what this "energy" may consist. However, they propose that it exerts a long-distance repulsive force. This dark energy is also known as Albert Einstein's cosmological constant," or a constant used by Einstein to complete his theory of gravitation. Einstein had originally called it his "greatest blunder," but this constant was resurrected in 1998 when astronomers discovered the recent acceleration of the expansion of the universe. If this is the case, the universe will continue to expand forever, long after all the stars have grown dark and cold.

IMPACT

WMAP data support the inflationary model of the universe and have provided the first accurate measure of the age of the universe. Not only did the data identify Hubble's constant to an extremely precise degree, but with the concomitantly precise calculation of the age of the universe many astronomers have come to accept the leading cosmological model: that the first few seconds after the big bang involved a very rapid, energetic expansion of space.

In order to confirm this model further, however, even more sensitive maps of the CMB are needed. Inflation should have created gravitational waves that would be imprinted on the CMB, and until these can be detected, other theories of early universe formation cannot be entirely ruled out.

See also Big Bang; Black Holes; Cosmic Microwave Background Radiation; Expanding Universe; Galaxies; Inflationary Model of the Universe; Quarks; String Theory.

FURTHER READING

National Aeronautics and Space Administration. "Wilkinson Microwave Anisotropy Probe Homepage." http://map.gsfc.nasa.gov.
Plionis, Manolis, ed. *Multiwavelength Cosmology*. New York: Springer, 2004.
Rowan-Robinson, Michael. *Cosmology*. London: Oxford University Press, 2003.

—Nicole E. Gugliucci

X RADIATION

THE SCIENCE: X rays, first observed by Wilhelm Röntgen, have remarkable penetrating power that has been widely used for medical and industrial applications.

THE SCIENTISTS:

Wilhelm Röntgen (1845-1923), experimental physicist who discovered X rays

William Crookes (1832-1919), inventor of the vacuum tube that bears his name

Heinrich Hertz (1857-1894), discoverer of radio waves

Philipp Lenard (1862-1947), experimental physicist who studied cathode rays

Ludwig Zehnder (1854-1935), Röntgen's laboratory assistant, colleague, and co-author of several research papers

CATHODE RAYS

X rays were discovered in 1895 by Wilhelm Röntgen, a professor of physics at the University of Würzburg, Germany. He was investigating the radiation produced in a partially evacuated glass bulb when a high voltage was applied.

William Crookes in 1869 had published a research report in which he described the bright glow that occurred inside such a bulb. Another physicist, Philipp Lenard, then showed that the electrical discharge inside the glass bulb could penetrate a thin aluminum window, producing an external beam that traveled through several centimeters of open air. This beam was called a "cathode ray" because it originated at the negative voltage terminal, which is the cathode. Lenard was able to trace the path of cathode rays by using fluorescent paint on a small screen that glowed in the dark when radiation struck it. He showed that cathode rays could be deflected by a magnet.

A MYSTERIOUS GLOW

Röntgen was an experienced experimentalist with twenty-five years of laboratory research and more than forty technical publications. Using the same type of apparatus as Crookes and Lenard, he first confirmed their observations for himself. In order to see the external beam more clearly, he surrounded the glass bulb with opaque, black paper, so that the light produced inside the bulb would be blocked out and the external beam would show up more clearly. On November 9, 1895, according to his laboratory notebook, he noticed something quite unusual. A piece of cardboard coated with fluorescent paint, lying on the table more than a meter away, started to glow whenever the electric discharge was turned on. This was a startling observation because cathode rays could not travel that far. Was there a new type of radiation coming through the black paper?

Working by himself, Röntgen began a systematic investigation of the mysterious

Wilhelm Röntgen. (Courtesy, Yale University)

rays. He observed fluorescence at a distance as much as two meters from the discharge tube. The radiation had penetrated opaque black paper, so he decided to test various other materials for their transparency. Even behind a book of one thousand pages, he found that the fluorescent screen lit up brightly. Blocks of wood and sheets of aluminum transmitted the radiation fairly well, but two millimeters of lead proved to be opaque. When holding his hand between the discharge apparatus and the fluorescent screen, he was able to see the shadow of the bones inside the faint outline of his fingers.

Photographs of the Invisible

Further experiments showed that photographic plates were sensitive to the radiation. This enabled Röntgen to make a permanent record of the observations that he had seen by eye previously. He had to be careful not to store unused photographic plates near the apparatus or they would become fogged by stray radiation. In his publications, Röntgen referred to the new type of radiation as X rays because they were a mystery. He used a glass prism to see if X rays could be refracted like ordinary light, but the result was negative. He also found that X rays were not reflected by a mirror and could not be focused by a lens. Diffraction gratings, which had been used to measure the wavelengths of visible light with high precision, had no effect on X rays, and a magnet caused no deflection.

On December 22, 1895, Röntgen asked his wife to help him in the laboratory. He placed the X-ray tube just underneath a table while she held her hand on the table surface with the photographic plate above her hand. The exposure time was about five minutes. When he developed the photograph, it showed the bones in her hand with her wedding ring on one finger. A photography assistant made multiple prints from this and several other negatives. On January 1, 1896, Röntgen sent a ten-page article with photographs to the Physical-Medical Society of Würzburg, as well as to colleagues at other universities. The pictures created a sensation. Nothing like them had ever been seen before.

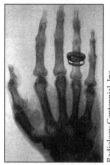

Radiology Centennial, Inc.

Within a few days, newspapers all over Europe had published stories and photographs about this new scientific development. A flood of messages came to Röntgen with invitations to give lectures and demonstrations. He turned them all down, except he could not refuse one that came from the emperor of Germany, Kaiser Wilhelm I. On January 13, he traveled to Berlin with his X-ray apparatus and showed to the assembled court how metal objects inside a closed box could be photographed. On January 23, he gave a lecture to the faculty and students at his own university. He told the audience about his experiments, giving credit to earlier contributions made by Hertz and Lenard. Toward the close of the lecture, he made an X-ray photograph of the hand of a faculty colleague, which was quickly developed and passed around the room. Prolonged applause came as the lecture ended.

Over the next several weeks, Röntgen received letters from many scientists who were experimenting with X rays. One person sent a photograph of a fish showing its detailed bone structure. His friend Ludwig Zehnder, whom he had known since graduate school in Zürich, took several photographs of the human body, which he pasted together to obtain a complete skeleton from head to foot. There were some crackpot letters, such as the one asking for a sum of money to solve the secrets of weather forecasting with X rays. The greatest honor for Röntgen was to be awarded the 1901 Nobel Prize, the first year in which the award was given.

Impact

Röntgen felt that the benefits of X rays should be available to humankind without restrictions. He did not take out a patent on his discovery even although it could have made him wealthy. His apparatus was not expensive or difficult to duplicate. The most difficult part would have been to get a glass blower to make a glass bulb with two metal electrodes inside, connected by wires going through the glass to two terminals on the outside. Many hospitals and research laboratories were able to set up their own X-ray machines. Within one year of Röntgen's initial publication, nearly one thousand articles on X rays appeared in various technical journals.

The medical profession enthusiastically welcomed X rays as a new diagnostic tool. Doctors were able to determine the severity of broken bones and to locate swallowed objects or an embedded bullet in the body. Annual chest X rays for schoolchildren became a routine procedure to diagnose early signs of tuberculosis. Irradiation of cancerous tumors was found to be a beneficial therapy as long as the dose was carefully regulated. In the 1970's, a major improvement in X-ray technology, called the CT-scan, was developed. A narrow beam of X rays was swept across a portion of the body from many different angles, and then the information was correlated by a computer to produce a picture on a screen.

X-ray apparatus came into common use at airports to inspect baggage before boarding. X rays have been used to search for hidden microphones in the wall of a room before a diplomatic conference. In the pipeline industry, after individual sections of pipe have been welded together, portable X-ray machines have been used to detect possible hairline cracks at the welds that might later allow fluid to leak. X-ray analysis has been widely used by chemists to determine the structure of complex molecules, such as the DNA helix. Röntgen's discovery of X rays, therefore, provides a fine example of how pure research can lead to a multitude of unanticipated practical applications.

See also Compton Effect; Electromagnetism; Electrons; Isotopes; Radioactivity; X Radiation; X-Ray Astronomy; X-Ray Crystallography; X-Ray Fluorescence.

FURTHER READING

Farmelo, Graham. "The Discovery of X-Rays." *Scientific American* (November, 1995): 86-91.

Hart, Michael H. *The 100: A Ranking of the Most Influential Persons in History.* New York: Galahad Books, 1982.

Kevles, Bettyann. *Naked to the Bones: Medical Imaging in the Twentieth Century.* Reading, Mass.: Addison Wesley, 1998.

Nitske, W. Robert. *The Life of Wilhelm Conrad Röntgen: Discoverer of the X Ray.* Tucson: University of Arizona Press, 1971.

Turner, G. L. "Röntgen, Wilhelm Conrad." In *Dictionary of Scientific Biography*, Volume 11: 529-531. New York: Charles Scribner's Sons, 1981.

Walker, James S. *Physics*. 2d ed. Upper Saddle River, N.J.: Pearson/ Prentice Hall, 2004.

—*Hans G. Graetzer*

X-Ray Astronomy

THE SCIENCE: Riccardo Giacconi and his colleagues launched a rocket-borne X-ray telescope that detected X rays from the constellation Scorpius.

THE SCIENTISTS:

Riccardo Giacconi (b. 1931), Italian American physicist and astronomer
Herbert Friedman (1916-2000), American cosmic-ray physicist
Norman Harmon (b. 1929), senior scientist at American Science and Engineering, Inc.
Frank Bethune McDonald (b. 1925), American astrophysicist

X RAYS FROM SCORPIUS

All stars in the universe emit electromagnetic radiation as a result of the enormous thermonuclear reactions and complex chemical reactions that take place within them. Such radiation comes in many forms that have their own frequencies and wavelengths. The electromagnetic spectrum ranges from low-frequency, long-wavelength radiations, such as radio, television, microwaves, and visible light, to higher-frequency, shorter-wavelength radiations such as ultraviolet rays, X rays, gamma rays, and cosmic rays. High-frequency radiations are mostly blocked by Earth's ozone layer, a chemical shield that reacts with these radiations as they bombard the Earth's atmosphere. Consequently, physicists and astronomers who wish to study extraterrestrial high-energy (high-frequency) radiations must place measuring instruments into orbit above the Earth's atmosphere.

During the late 1950's, a group of physicists at the Massachusetts Institute of Technology (MIT) in Cambridge, Massachusetts, established a company whose primary focus was high-energy physics and space research. The company, American Science and Engineering, Inc., joined forces with Riccardo Giacconi of Princeton University in 1959 to establish a space science research division.

In collaboration with the National Aeronautics and Space Administration (NASA), American Science and Engineering decided to test the emission of X rays from stars. This work had been started in the late 1940's by Herbert Friedman of the Naval Research Laboratory. Friedman launched X-ray detectors above the Earth's atmosphere aboard captured German V-2 rockets and demonstrated that the Sun emits X rays.

During 1960 and 1961, a research team that included Giacconi and Norman Harmon devised a small, highly sensitive X-ray telescope that could detect faint X-ray emissions from specific regions of space and that could ride aboard rockets that could fly as high as 160 kilometers. They attempted several launches of X-ray telescopes from White Sands Missile Range in New Mexico beginning in the fall of 1961. At midnight on June 18-19, 1962, with a six-minute suborbital flight, the X-ray telescope received and recorded stellar X-ray emissions on film.

Careful analysis of the X-ray film showed a higher emission of X rays emanating from the southern constellation Scorpius. They named this X-ray source Scorpius X-1, the first-discovered X-ray source outside the solar system. Friedman's research group quickly confirmed Giacconi's discovery. Other high-energy astrophysicists entered the field and discovered additional X-ray sources, including Cygnus X-1 and the Crab nebula in Taurus. The ultimate goal for Giacconi and his colleagues was to place a series of orbiting X-ray telescopes around Earth for precise measurements of hundreds of stellar X-ray sources. They planned the development of these satellites in coordination with Frank Bethune McDonald of the NASA Goddard Space Flight Center in Greenbelt, Maryland.

Launching Uhuru

During the late 1960's, Giacconi and his colleagues continued their work on a proposed orbiting X-ray telescope. Friedman's group at the Naval Research Laboratory and McDonald's group at the Goddard Space Flight Center were working toward the same goal. On December 12, 1970, the Cosmic X-Ray Explorer satellite was launched from an oil rig located off the coast of Kenya. Kenya was chosen because from there the satellite could easily enter an orbit that would carry the satellite around the Earth's equator, enabling the X-ray telescope to detect X-ray sources from practically every direction around Earth. X-ray data were relayed to a ground-based control station. The X-Ray Explorer, the Small Astronomy Satellite 1, was nicknamed Uhuru, the Kenyan word for "freedom," because it was launched on Kenya's Independence Day.

Many X-ray sources were identified by Uhuru as sunlike stars and galactic nuclei. Still other sources were determined to be superdense collapsed stars called neutron stars. Other scientists speculate that some X-ray sources (Cygnus X-1) may be black holes, gravitational singularities that are collapsed stars so dense that matter and light cannot escape.

LATER X-RAY OBSERVATORIES

Additional Small Astronomy Satellites were launched during the early 1970's, each satellite carrying a variety of high-energy detection equipment designed by Giacconi, Friedman, McDonald, and other physicists. A more advanced satellite series, the High-Energy Astronomy Observatory, consisted of three satellites. It was created in 1978 and was operational through 1981. It contained a powerful X-ray telescope that detected X-ray emissions from "quasars" (quasi-stellar radio sources), the most distant and oldest objects yet discovered in the universe. Other satellites followed: the Einstein Observatory, in 1979, the first satellite with focusing X-ray mirrors enabling it to see fainter sources; the Röntgen X-Ray Satellite (ROSAT) in 1990—a joint project of Germany, the United Kingdom, and the United States—the first satellite to make an all-sky survey with an imaging telescope; and the Japanese satellite ASCA, launched in 1993 by Japan and the United States, the first to use the new-generation charge-coupled devices (CCD) x-ray detectors.

One of the most ambitious projects in X-ray astronomy was launched on July 23, 1999, when the Chandra X-Ray Observatory (CXO) was lofted into orbit aboard space shuttle *Columbia*. This observatory is designed to image and measure the temperatures of extremely hot objects such as supernova remnants, neutron star accretion disks, and cosmic gas clouds. It is far more sensitive than previous x-ray telescopes and capable of revealing much finer detail.

The XMM-Newton was successfully launched December 10, 1999. It can detect fainter sources than Chandra, but Chandra has better resolution, enabling it to record finer details.

DATA FROM CHANDRA

Chandra's results often confirm what previously was only suspected, but occasionally it has made completely unanticipated discoveries as well. The facts that the Milky Way's black hole candidate emits fewer X rays than expected and that the environment of Andromeda's black hole candidate is cooler than expected strongly suggest that the processes involved are more complex and less well understood than previously supposed.

To see what faint sources might be present, scientists pointed Chandra at a small patch of sky in the direction of the constellation Canis Venatici and collected data for 27.7 hours. Since the early 1960's scientists have known that space is filled with a faint X-ray glow, but they did not know if that glow came from very hot diffuse gas spread throughout the universe or if it came from a large number of discrete sources. ROSAT had previously shown that much of the lower-energy X-ray background comes from distant objects such as quasars or active galactic nuclei (AGNs). AGNs are thought to be supermassive (billion-solar-mass) black hole candidates that are rapidly accreting more mass. Mass spiraling inward forms an accretion disk about the black hole candidate, and the gravitational energy released heats the disk so that it emits gamma rays, X rays, and visible light.

With better resolution and sensitivity, Chandra confirmed the ROSAT result and extended it to higher-energy X rays. Most of the X-ray background does come from discrete sources. Chandra found that about one-third of the sources are AGNs with brightly shining cores, but that another third of the X-ray sources are galactic nuclei that emit little or no visible light from their cores. Perhaps dust or gas surrounding their cores blocks visible light. If so, there may be tens of millions of similar objects over the whole sky, and the optical surveys of AGNs are very incomplete. Chandra found that the final third of the X-ray sources are in ultrafaint galaxies, galaxies that are barely detectable, if at all, in visible light. If they are so faint because they are far away, they would be among the most distant objects ever discovered.

Impact

Giacconi's discovery of the first extrasolar X-ray source was a tremendous astronomical achievement that changed scientists' view of the universe and led to a greater understanding of stellar astrophysics. The knowledge that many objects, including stars, planets, galaxies, and quasars, emit X rays has enabled scientists to comprehend the nature of these objects and the processes that occur within them. Giacconi's discovery created the field of X-ray astronomy, which continually yields new information about the universe.

The first X-ray telescopes, launched aboard sounding rockets by Giacconi and Friedman, pioneered later missions that revealed many cosmic X-ray emitters. With succeeding X-ray telescope missions, X-ray sources were discovered in every section of the universe. Soon, astronomers were able to draw up a comprehensive map of stellar X-ray emission.

In a larger context, X-ray astronomy is part of a larger movement, starting in the 1930's with radio astronomy, to use the nonvisible portions of the electromagnetic spectrum—from radio waves and microwaves to ultraviolet ranges—to gather data on the universe and its objects. What is "visible" in these ranges has provided more insight into the universe and its dynamics than the preceding four hundred years of visible telescopy.

See also Big Bang; Black Holes; Cepheid Variables; Chandrasekhar Limit; Cosmic Microwave Background Radiation; Cosmic Rays; Electromagnetism; Expanding Universe; Extrasolar Planets; Galactic Superclusters; Galaxies; Gamma-Ray Bursts; Hubble Space Telescope; Inflationary Model of the Universe; Neutron Stars; Ozone Hole; Pulsars; Quasars; Radio Astronomy; Radio Galaxies; Radio Maps of the Universe; Solar Wind; Space Shuttle; Spectroscopy; Stellar Evolution; Very Long Baseline Interferometry; Wilkinson Microwave Anisotropy Probe; X Radiation.

FURTHER READING

Bartusiak, Marcia. *Thursday's Universe*. New York: Times Books, 1986.
Cowen, R. "X-Ray Data Reveal Black Holes Galore." *Science News*, January 15, 2000, 36.
Fabian, A. C., K. A. Pounds, and R. D. Blandford. *Frontiers of X-Ray Astronomy*. London: Cambridge University Press, 2004.
Schlegel, Eric M. *The Restless Universe: Understanding X-Ray Astronomy in the Age of Chandra and Newton*. London: Oxford University Press, 2002.
Tucker, Wallace H. and Karen Tucker. *Revealing the Universe: The Making of the Chandra X-Ray Observatory*. Cambridge, Mass.: Harvard University Press, 2001.

—*David Wason Hollar, Jr., and David G. Fisher*

X-RAY CRYSTALLOGRAPHY

THE SCIENCE: The invention of X-ray crystallography provided an important technique for using X rays to determine the crystal structures of many substances.

THE SCIENTISTS:
Sir William Henry Bragg (1862-1942), English mathematician and physicist and cowinner of the 1915 Nobel Prize in Physics
Sir Lawrence Bragg (1890-1971), the son of Sir William Henry Bragg and cowinner of the 1915 Nobel Prize in Physics

Max von Laue (1879-1960), German physicist who won the 1914 Nobel
Prize in Physics
Wilhelm Conrad Röntgen (1845-1923), German physicist who won the
1901 Nobel Prize in Physics
René-Just Haüy (1743-1822), French mathematician and mineralogist
Auguste Bravais (1811-1863), French physicist

CRYSTALS

A crystal is a body that is formed once a chemical substance has solidi-
fied. It is uniformly shaped, with angles and flat surfaces that form a network based on the internal structure of the crystal's atoms. Determining
what these internal crystal structures look like is the goal of the science of
X-ray crystallography. To do this, it studies the precise arrangements into
which the atoms are assembled.

Central to this study is the principle of X-ray diffraction. This technique
involves the deliberate scattering of X rays as they are shot through a crystal, an act that interferes with their normal path of movement. The way in
which the atoms are spaced and arranged in the crystal determines how
these X rays are reflected off them while passing through the material. The
light waves thus reflected form a telltale interference pattern. By studying
this pattern, scientists can discover variations in the crystal structure.

The development of X-ray crystallography in the early twentieth century helped to answer two major scientific questions: What are X rays? and
What are crystals? It gave birth to a new technology for the identification
and classification of crystalline substances.

From studies of large, natural crystals, chemists and geologists had established the elements of symmetry through which one could classify, describe, and distinguish various crystal shapes. René-Just Haüy, about a
century before, had demonstrated that diverse shapes of crystals could be
produced by the repetitive stacking of tiny solid cubes.

Auguste Bravais later showed, through mathematics, that all crystal
forms could be built from a repetitive stacking of three-dimensional arrangements of points (lattice points) into "space lattices," but no one had
ever been able to prove that matter really was arranged in space lattices.
Scientists did not know if the tiny building blocks modeled by space lattices actually were solid matter throughout, like Haüy's cubes, or if they
were mostly empty space, with solid matter located only at the lattice
points described by Bravais.

With the disclosure of the atomic model of Danish physicist Niels Bohr
in 1913, determining the nature of the building blocks of crystals took on a

special importance. If crystal structure could be shown to consist of atoms at lattice points, then the Bohr model would be supported, and science then could abandon the theory that matter was totally solid.

X Rays Explain Crystal Structure

In 1912, Max von Laue first used X rays to study crystalline matter. Laue had the idea that irradiating a crystal with X rays might cause diffraction. He tested this idea and found that X rays were scattered by the crystals in various directions, revealing on a photographic plate a pattern of spots that depended on the orientation and the symmetry of the crystal.

The experiment confirmed in one stroke that crystals were not solid and that their matter consisted of atoms occupying lattice sites with substantial space in between. Further, the atomic arrangements of crystals could serve to diffract light rays. Laue received the 1914 Nobel Prize in Physics for his discovery of the diffraction of X rays in crystals.

Still, the diffraction of X rays was not yet a proved scientific fact. Sir William Henry Bragg contributed the final proof by passing one of the diffracted beams through a gas and achieving ionization of the gas, the same effect that true X rays would have caused. He also used the spectrometer he built for this purpose to detect and measure specific wavelengths of X rays and to note which orientations of crystals produced the strongest reflections. He noted that X rays, like visible light, occupy a definite part of the electromagnetic spectrum. Yet most of Bragg's work focused on actually using X rays to deduce crystal structures.

Bragg's son, Sir Lawrence Bragg, was also deeply interested in this new phenomenon. In 1912, he had the idea that the pattern of spots was an indication that the X rays were being reflected from the planes of atoms in the crystal. If that were true, Laue pictures could be used to obtain information about the structures of crystals. Bragg developed an equation that described the angles at which X rays would

William Henry Bragg. (The Nobel Foundation)

be most effectively diffracted by a crystal. This was the start of the X-ray analysis of crystals.

William Henry Bragg had at first used his spectrometer to try to determine whether X rays had a particulate nature. It soon became evident, however, that the device was a far more powerful way of analyzing crystals than the Laue photograph method had been. Not long afterward, father and son joined forces and founded the new science of X-ray crystallography. By experimenting with this technique, Lawrence Bragg came to believe that if the lattice models of Bravais applied to actual crystals, a crystal structure could be viewed as being composed of atoms arranged in a pattern consisting of a few sets of flat, regularly spaced, parallel planes.

Diffraction became the means by which the Braggs deduced the detailed structures of many crystals. Based on these findings, they built three-dimensional scale models out of wire and spheres that made it possible for the nature of crystal structures to be visualized clearly even by nonscientists. Their results were published in the book *X-Rays and Crystal Structure* (1915).

IMPACT

The Braggs founded an entirely new discipline, X-ray crystallography, which continues to grow in scope and application. Of particular importance was the early discovery that atoms, rather than molecules, determine the nature of crystals. X-ray spectrometers of the type developed by the Braggs were used by other scientists to gain insights into the nature of the atom, particularly the innermost electron shells. The tool made possible the timely validation of some of Bohr's major concepts about the atom.

X-ray diffraction became a cornerstone of the science of mineralogy. The Braggs, chemists such as Linus Pauling, and a number of mineralogists used the tool to do pioneering work in deducing the structures of all major mineral groups. X-ray diffraction became the definitive method of identifying crystalline materials.

Metallurgy progressed from a technology to a science as metallurgists became able, for the first time, to deduce the structural order of various alloys at the atomic level.

Diffracted X rays were also applied in the field of biology, particularly at the Cavendish Laboratory under the direction of Lawrence Bragg. X-ray crystallography proved to be essential for deducing the structures of hemoglobin, proteins, viruses, and eventually the double-helix structure of deoxyribonucleic acid (DNA).

See also X-Ray Fluorescence.

FURTHER READING

Achilladelis, Basil, and Mary Ellen Bowden. *Structures of Life*. Philadelphia: The Center, 1989.

Bragg, William Lawrence. *The Development of X-Ray Analysis*. New York: Hafner Press, 1975.

Thomas, John Meurig. "Architecture of the Invisible." *Nature* 364 (August 5, 1993).

—*Edward B. Nuhfer*

X-RAY FLUORESCENCE

THE SCIENCE: By studying the interaction between X rays and matter, Charles Glover Barkla succeeded in determining important physical characteristics of X rays and the atomic structure of matter.

THE SCIENTISTS:

Charles Glover Barkla (1877-1944), English physicist who was awarded the 1917 Nobel Prize in Physics for his work on X-ray scattering and his discovery of the characteristic Röntgen radiations of the elements

Wilhelm Conrad Röntgen (1845-1923), German physicist who discovered X rays in 1895 and was a recipient of the first Nobel Prize in Physics in 1901

George Gabriel Stokes (1819-1903), eminent British mathematician and physicist who theorized about the cause and nature of X rays

Sir Joseph John Thomson (1856-1940), British physicist and teacher who startled scientists by announcing his experimental confirmation that cathode rays consisted of charged particles more than one thousand times lighter than the smallest atom and was awarded the 1906 Nobel Prize in Physics

MATTER IN A NEW STATE

For several decades in the nineteenth century, physicists studied the cathode rays. For an even longer time, scientists had known of the existence of atoms. Nevertheless, during the last decade of the nineteenth century, scientists still had great difficulty in comprehending the physical nature of either cathode rays or atoms. Apparently, atoms of some chemical elements were heavier, and others were lighter. It was not known why. The reason could have been that atoms consisted of different materials, or that the heavier ones had more of the same materials. Chemical facts gave clues about the existence of atoms. Cathode rays appeared to be "tentacles" originating from the atom.

In December, 1895, a sequence of clues, and tentacles, emerged. First, Wilhelm Conrad Röntgen reported from the University of Würzburg that he had discovered X rays. In 1896, Antoine-Henri Becquerel announced to the French Academy of Sciences that uranium spontaneously emitted invisible radiations, which would blacken a photographic plate yet which seemed different from X rays. In 1898, Pierre Curie and Marie Curie detected two new elements that apparently also emitted the same types of radiation. At the same time, in England, Sir Joseph John Thomson made remarkable progress in the study of the cathode rays. He would conclude from experimental evidence that these rays consisted of charged particles. He declared that these charged particles were "matter in a new state" and that the chemical elements were made up of matter.

X-Ray Scattering

By the beginning of the twentieth century, scientists were confronted with a number of intriguing questions about the nature of X rays, how to account for radioactivity, and how to reconcile the apparent endlessness of radioactive emanations with the principle of the first law of thermodynamics, the conservation of energy. It was in this atmosphere of challenging scientific inquiry that Charles Glover Barkla began his scientific career. From 1899 to 1902, he conducted research with Thomson. In 1902, he attended University College in Liverpool and began his lifelong study of X rays.

The veteran mathematical physicist George Gabriel Stokes proposed the "ether pulse" theory about the nature of X rays. He hypothesized that X rays were irregular electromagnetic pulses created by the irregular accelerations of the cathode rays when they were stopped by the atoms in the target of the X-ray tube. With this theory, Thomson derived a mathematical formula expressing the scattering of X rays by electrons. Barkla's first research project was to test experimentally Stokes's and Thomson's theories.

Five years previously, Georges Sagnac had experimented in France on the absorption of X rays by solids—a phenomenon that was directly related to scattering. Sagnac found that the secondary scattered radiation was of distinctly greater absorbability. Barkla showed that the secondary radiation from light gaseous elements was of the same absorbability as that of the primary beam. He worked on air first, then extended the investigation to hydrogen, carbon dioxide, sulfur dioxide, and hydrogen sulfide. The presence of the secondary radiation was tested by an electroscope, with the assumption that the amount of ionization should be proportional to the intensity of the radiation passing through the instrument.

To check the absorbability of the rays—primary and secondary—Barkla used a thin aluminum plate. He published the results in 1903. At the time, the fact that scattering did not modify the absorbability of the radiation appeared to be strong support for the ether pulse theory. From the same set of experiments, Barkla demonstrated that "this scattering is proportional to the mass of the atom." This was a highly satisfying result because it supported the theory that the atoms of different substances are different systems of similar corpuscles, where, in the atom, the number is proportional to its atomic weight. These similar corpuscles, according to most contemporary physicists, was Thomson's "matter in a new state."

Polarized X Rays

In 1904, Barkla began a new series of experiments that would disclose additional physical characteristics of X rays. Because ordinary light, as the propagation of electromagnetic oscillations, is a transverse wave, it can be polarized relatively easily: When it is scattered in the direction at right angles to the incident (primary) beam, the transverse vibrations constituting the light are confined to a plane perpendicular to the primary beam. Barkla was researching the question of whether X rays could be polarized in the same way so that they could be confirmed as electromagnetic waves. This proved to be a serious challenge. It took Barkla two years to perform the difficult experiment and arrive at a clear conclusion that the scattered beam was highly polarized; consequently, X rays were most probably transverse waves, like ordinary light.

X-Ray Scattering

While investigating the intensity in different directions of the secondary radiation, Barkla found that light elements—such as carbon, aluminum, and sulfur—showed marked variation in intensity with direction; calcium showed much less. With iron and even heavier elements, there was practically no difference in intensity in different directions. This phenomenon led Barkla to a closer investigation of the relation between atomic weight and absorbability. The result of his experiments showed that for light elements, the scattered radiation closely resembled the primary radiation, but for elements heavier than calcium, the scattered radiation was quite different from the primary. When Barkla examined the scattered (secondary) radiation more closely, he found that the secondary radiation from metals contained not only scattered radiation of the same character as the primary but also homogeneous radiation that was characteristic of the metallic element itself.

Atomic Weight and Characteristic Radiation

Meanwhile, Barkla also discovered an X-ray phenomenon that was analogous to a discovery made by Stokes: Fluorescent substances fluoresced only when exposed to light of shorter wavelength than that of the fluorescent light emitted by the substance. This phenomenon is known as Stokes's law. Barkla found that the emission of the homogeneous (secondary) radiation occurred only when the incident X-ray beam was harder than the characteristic radiation itself. Moreover, Barkla found some revealing facts about the homogeneous characteristic radiations. Beginning with calcium, and moving toward the heavier elements, the characteristic X-ray radiations form one or two series. From calcium (atomic weight 40) to rhodium (atomic weight 103), there appeared a K series; from silver (atomic weight 108) to cerium (atomic weight 140), there appeared a K series and an L series; from tungsten (atomic weight 184) to bismuth (atomic weight 208), there appeared an L series only. K radiations were softer, L radiations harder. The heavier the atom was, the harder its characteristic radiations. Such phenomena, correlating atomic weight to characteristic X rays closely, showed that the latter must have originated from the atom. In fact, Barkla's discoveries anticipated the assignment to each chemical element an atomic number, which, in general, was recognized as about one-half the atomic weight.

Impact

Following these discoveries in 1906, Barkla and other physicists researched interactions between X rays and matter and achieved historic results. These achievements, accomplished between 1909 and 1923, were categorized in three stages.

First, X rays interact with crystal lattices. In 1909, Max von Laue attended the University of Munich and was influenced by Röntgen and mineralogists who informed him of theories on the structure of crystal solids. Von Laue proceeded to combine the study of X rays with that of solid structures. He developed a mathematical theory based on the assumption that crystal lattices could serve as "diffraction gratings" (a type of instrument in optical experiments that demonstrates the wave character of light) for X rays. This idea was experimentally confirmed, and the confirmation has been highly praised. It opened vast potentials for studying the nature of X rays and the structure of crystal solids. Shortly after von Laue's publication of his work in 1912, William Henry Bragg and his eldest son, Lawrence Bragg, founded the science of crystallography. In particular, William Henry Bragg created the "ionization spectrometer" for measuring the exact wavelengths of X rays; Lawrence Bragg derived the influential equation now named after him. The Bragg equation tells at what angles X rays will be most efficiently diffracted by a crystal layer.

The second stage was the recognition that X rays interact with atoms, especially heavy atoms. Henry Moseley used the Bragg spectrometer soon after its introduction to study the characteristic X rays from the atom. With a new, powerful instrument, Moseley turned to Barkla's line of investigation. Moseley could now measure Barkla's K series and L series with exactness. Significantly extending such measurements, he made wonderful discoveries that have since been called Moseley's law: the mathematical formula that relates the X-ray spectrum of an element to its atomic number. Moseley also made a series of verifiable predictions about the periodic table of elements. Tragically, Moseley was killed in World War I at the Dardanelles. Later, studies in X-ray spectroscopy and its interpretation were accomplished by Karl Manne Georg Siegbahn, winner of the 1924 Nobel Prize in Physics.

The third stage was the recognition that X rays interact with light atoms (free electrons). When Barkla delivered his Nobel Prize speech in 1920 (for Physics in 1917), he declared that in the phenomena of scattering, there is strong positive evidence against any quantum theory. Three years later, in 1923, Arthur Holly Compton was to prove the folly of Barkla's statement. He followed Barkla in experimenting on the comparison of secondary X rays with primary X rays, especially when the former were scattered from light atoms. Compton experimented with the spectrometer and theoretically with the concept of photons and was awarded the Nobel Prize in 1927.

See also X-Ray Crystallography.

FURTHER READING

Allen, H. S. "Charles Glover Barkla, 1877-1944." *Obituary Notices of Fellows of the Royal Society of London* 5 (1947): 341-366.
Forman, Paul. "Charles Glover Barkla." In *Dictionary of Scientific Biography*, edited by C. C. Gillispie. New York: Charles Scribner's Sons, 1970.

Heathcote, Niels Hugh de Vaudrey. *Nobel Prize Winners in Physics, 1901-1950.* New York: Henry Schuman, 1953.

—*Wen-yuan Qian*

YELLOW FEVER VACCINE

THE SCIENCE: The first safe vaccine against the virulent yellow fever virus mitigated some of the deadliest epidemics of the nineteenth and early twentieth centuries.

THE SCIENTISTS:
Max Theiler (1899-1972), South African microbiologist
Wilbur Augustus Sawyer (1879-1951), American physician
Hugh Smith (1902-1995), American physician

A YELLOW FLAG

Yellow fever, caused by a virus and transmitted by mosquitoes, infects humans and monkeys. After the bite of the infecting mosquito, it takes several days before symptoms appear. The onset of symptoms is abrupt, with headache, nausea, and vomiting. Because the virus destroys liver cells, yellowing of the skin and eyes is common. Approximately 10 to 15 percent of patients die after exhibiting the terrifying signs and symptoms. Death occurs usually from liver necrosis (decay) and liver shutdown. Those that survive, however, recover completely and are immunized.

At the beginning of the twentieth century, there was no cure for yellow fever. The best that medical authorities could do was to quarantine the afflicted. Those quarantines usually waved the warning yellow flag, which gave the disease its colloquial name, "yellow jack."

After the *Aëdes aegypti* mosquito was clearly identified as the carrier of the disease in 1900, efforts were made to combat the disease by wiping out the mosquito. Most famous in these efforts were the American army surgeon Walter Reed and the Cuban physician Carlos J. Finlay. This strategy was successful in Panama and Cuba and made possible the construction of the Panama Canal. Still, the yellow fever virus persisted in the tropics, and the opening of the Panama Canal increased the danger of its spreading aboard the ships using this new route.

Moreover, the disease, which was thought to be limited to the jungles of South and Central America, had begun to spread arounds the world to wherever the mosquito *Aëdes aegypti* could carry the virus. Mosquito larvae traveled well in casks of water aboard trading vessels and spread the disease to North America and Europe.

Immunization by Mutation

Max Theiler received his medical education in London. Following that, he completed a four-month course at the London School of Hygiene and Tropical Medicine, after which he was invited to come to the United States to work in the department of tropical medicine at Harvard University.

Max Theiler. (The Nobel Foundation)

While there, Theiler started working to identify the yellow fever organism. The first problem he faced was finding a suitable laboratory animal that could be infected with yellow fever. Until that time, the only animal successfully infected with yellow fever was the rhesus monkey, which was expensive and difficult to care for under laboratory conditions. Theiler succeeded in infecting laboratory mice with the disease by injecting the virus directly into their brains.

Laboratory work for investigators and assistants coming in contact with the yellow fever virus was extremely dangerous. At least six of the scientists at the Yellow Fever Laboratory at the Rockefeller Institute died of the disease, and many other workers were infected. In 1929, Theiler was infected with yellow fever; fortunately, the attack was so mild that he recovered quickly and resumed his work.

During one set of experiments, Theiler produced successive generations of the virus. First, he took virus from a monkey that had died of yellow fever and used it to infect a mouse. Next, he extracted the virus from that mouse and injected it into a second mouse, repeating the same procedure using a third mouse. All of them died of encephalitis (inflammation of the brain). The virus from the third mouse was then used to infect a monkey. Although the monkey showed signs of yellow fever, it recovered completely. When Theiler passed the virus through more mice and then into the abdomen of another monkey, the monkey showed no symptoms of the disease. The results of these experiments were published by Theiler in the journal *Science*.

This article caught the attention of Wilbur Augustus Sawyer, director of the Yellow Fever Laboratory at the Rockefeller Foundation International Health Division in New York. Sawyer, who was working on a yellow fever vaccine, offered Theiler a job at the Rockefeller Foundation, which Theiler accepted. Theiler's mouse-adapted, "attenuated" virus was given to the laboratory workers, along with human immune serum, to protect them against the yellow fever virus. This type of vaccination, however, carried the risk of transferring other diseases, such as hepatitis, in the human serum.

In 1930, Theiler worked with Eugen Haagen, a German bacteriologist, at the Rockefeller Foundation. The strategy of the Rockefeller laboratory was a cautious, slow, and steady effort to culture a strain of the virus so mild as to be harmless to a human but strong enough to confer a long-lasting immunity. (To "culture" something—tissue cells, microorganisms, or other living matter—is to grow it in a specially prepared medium under laboratory conditions.) They started with a new strain of yellow fever harvested from a twenty-eight-year-old West African named Asibi; it was later known as the "Asibi strain." It was a highly virulent strain that in four to seven days killed almost all the monkeys that were infected with it. From time to time, Theiler or his assistant would test the culture on a monkey and note the speed with which it died.

It was not until April, 1936, that Hugh Smith, Theiler's assistant, called to his attention an odd development as noted in the laboratory records of strain 17D. In its 176th culture, 17D had failed to kill the test mice. Some had been paralyzed, but even these eventually recovered. Two monkeys who had received a dose of 17D in their brains survived a mild attack of encephalitis, but those who had taken the infection in the abdomen showed no ill effects whatever. Oddly, subsequent subcultures of the strain killed monkeys and mice at the usual rate. The only explanation possible was that a mutation had occurred unnoticed.

The batch of strain 17D was tried over and over again on monkeys with no harmful effects. Instead, the animals were immunized effectively. Then it was tried on the laboratory staff, including Theiler and his wife, Lillian. The batch injected into humans had the same immunizing effect. Neither Theiler nor anyone else could explain how the mutation of the virus had resulted. Attempts to duplicate the experiment, using the same Asibi virus, failed. Still, this was the first safe vaccine for yellow fever. In June, 1937, Theiler reported this crucial finding in the *Journal of Experimental Medicine*.

Impact

Following the discovery of the vaccine, Theiler's laboratory became a production plant for the 17D virus. Before World War II (1939-1945), more than one million vaccination doses were sent to Brazil and other South American countries. After the United States entered the war, eight million soldiers were given the vaccine before being shipped to tropical war zones. In all, approximately fifty million people were vaccinated in the war years.

Yet although the vaccine, combined with effective mosquito control, eradicated the disease from urban centers, yellow fever is still present in large regions of South and Central America and of Africa. The most severe outbreak of yellow fever ever known occurred from 1960 to 1962 in Ethiopia; out of one hundred thousand people infected, thirty thousand died.

The 17D yellow fever vaccine prepared by Theiler in 1937 continues to be the only vaccine used by the World Health Organization, more than fifty years after its discovery. There is a continuous effort by that organization to prevent infection by immunizing the people living in tropical zones.

See also AIDS; Human Immunodeficiency Virus; Hybridomas; Immunology; Oncogenes; Polio Vaccine: Sabin; Polio Vaccine: Salk; Smallpox Vaccination; Stem Cells; Streptomycin; Viruses.

Further Reading

DeJauregui, Ruth. *One Hundred Medical Milestones That Shaped World History*. San Mateo, Calif.: Bluewood Books, 1998.
Delaporte, François. *The History of Yellow Fever: An Essay on the Birth of Tropical Medicine*. Cambridge, Mass.: MIT Press, 1991.
Theiler, Max, and Wilbur G. Downs. *The Arthropod-borne Viruses of Vertebrates: An Account of the Rockefeller Foundation Virus Program, 1951-1970*. New Haven, Conn.: Yale University Press, 1973.
Williams, Greer. *Virus Hunters*. London: Hutchinson, 1960.

—*Gershon B. Grunfeld*

Zinjanthropus

The Science: At Olduvai Gorge, Tanzania, Louis and Mary Leakey discovered *Zinjanthropus boisei* (later reclassified as *Australopethicus boisei*), one of the oldest hominid fossils.

THE SCIENTISTS:
Louis S. B. Leakey (1903-1972), English anthropologist
Mary Leakey (1913-1996), English archaeologist and anthropologist

OLDUVAI FOSSIL BEDS

The Olduvai Gorge in northern Tanzania owes its origins to massive geological faulting approximately 100,000 years ago. As a result of the changing geology, the Great Rift Valley was formed, which stretches over 6,400 kilometers of East Africa, from Jordan in the north through Kenya and Tanzania to Mozambique in the south. A newly formed river rapidly cut through the previously laid down strata. The strata were formed from a series of Ngorongoro and Lemagrut volcanic eruptions, combined with lake and river deposits laid down millions of years ago.

The gorge has four distinct layers, or beds, numbered I (at the bottom of the gorge) through IV (nearest the top). Bed I is the oldest and has been dated to be more than 2 million years old. While Olduvai is more than 40 kilometers in length and approximately 92 meters deep, it is a small portion of the Great Rift system.

A stone-tool technology was encountered at Olduvai, along with the discovery of several extinct vertebrates, including a 26-million-year-old primate, a member of the genus *Proconsul*. The Oldowan tool tradition, named for Olduvai, was dated to the Lower Pleistocene epoch, which, at that time, was believed to have begun about a million years ago. (More recent dating techniques have pushed this age back another million years.)

THE NUTCRACKER MAN

Louis and Mary Leakey, both anthropologists, had been introduced to the Oldowan tradition and were conducting research in the area, Louis beginning in 1931 and Mary in 1935. On the morning of July 17, 1959, Mary discovered the remains for which she and her husband had long been searching: the animal believed to be responsible for the previously discovered Oldowan tools. She had happened upon the upper dentition and a few fragments of a never-before-documented hominid fossil. The fossil was found very near the bottom of the gorge.

During the next nineteen days, the Leakeys recovered more than four hundred pieces from an almost complete skull. Similar hominid fossils (later reclassified as members of the genus *Australopithecus*) had been found previously in South Africa by anthropologist Raymond Arthur Dart in 1924 and paleontologist Robert Broom in 1936. Yet firm dates could not be established for the South African finds; evidence of associated tool use was not as accurately documented as that encountered at Olduvai.

The hominid discovered by the Leakeys is thought to have lived approximately 1.75 million years ago. They recognized the remains as those of a young adult male, basing their conclusion on the degree of dental

LOUIS AND MARY LEAKEY

Louis S. B. Leakey, born to English missionary parents in Kenya and initiated into the Kikuyu tribe as a young boy, had varied interests but was ultimately trained as an anthropologist at the University of Cambridge. In 1931, he was accompanied on his first paleontological expedition to Olduvai by Hans Reck, a German geologist. Reck, who had worked at Olduvai prior to 1914, discouraged Leakey from his hope of finding evidence of prehistoric human activity at the gorge; however, within the first day of their arrival, a hand ax was discovered. Leakey recognized this site as an important one, and it was to become famous twenty-eight years later with the discovery by Mary Leakey, his second wife, of the hominid fossil *Zinjanthropus boisei*.

Mary Douglas Nicol was born in England in 1913. She was educated in England as an archaeologist. When she met Leakey in 1933, she was becoming well known for her illustrations of lithic tools. Indeed, it was soon after they met that Leakey asked her to undertake the drawings for the first edition of his book *Adam's Ancestors* (1934). They were married on Christmas Eve, 1936, in England, only days before their departure for Kenya. In 1959, Mary found the *Zinjanthropus* (*Australopithecus boisei*) fossil, which was to propel the Leakey family to worldwide fame. Louis Leakey became something of an instant celebrity with this discovery (if Mary did not), with his picture on newspaper and magazine covers along with the "Nutcracker man."

Mary and Louis grew apart after the *boisei* discovery. By consensus the better scientist of the two, Mary remained at Olduvai working alone for the most part. Louis pursued other projects, as well as some of the benefits of fame. In 1974, Mary began working at Laetoli to the south, where in 1978 she and her colleagues found the amazingly complete footprints of two bipedal hominids cast in volcanic ash. She continued working with paleoanthropologist Tim White, who had also worked with Donald Johanson of Lucy fame. She and White discovered hominid bones and more footprints—narrow and arched, much like those of modern humans—that dated between 3.7 and 3.5 million years ago, about twice the age of *boisei*.

In 1983, Mary retired to Nairobi after more than two decades in the field. She died thirteen years later, at age eighty-three. Despite her lack of formal education, she is considered one of the top anthropologists of the twentieth century.

eruption and development and the evidence of extreme robustness. Furthermore, the dental, facial, and cranial morphology (shape and structure) of the Leakey discovery was distinct from the hominids previously known from South Africa. As a result, the Leakeys classified their discovery into a new genus, *Zinjanthropus*, and species, *boisei*. *Zinj* is Arabic for "East Africa," *anthropus* is Greek for "humankind," and *boisei* is a latinization of "Boise," the family name of Leakey's benefactor, Charles Boise. Because of the specimen's cranial robustness and massive teeth, the fossil's popular name became Nutcracker man.

IMPACT

The discovery of *Zinjanthropus* affected human paleontology in many ways. The age of the first hominids was pushed back dramatically. Although Dart, Broom, and others had previously given the world cause to accept the notion proposed by the English naturalist Charles Darwin in 1871 that Africa was the cradle of humankind, the various hominid fossils recovered from South Africa did not lend themselves to accurate dating. The discovery of *Zinjanthropus boisei* and the Oldowan tools from the volcanic contexts of Olduvai Gorge, however, allowed accurate radiometric dates

The Leakeys (from left): Son Richard, Mary, and Louis. (Win Parks, National Geographic Society)

to be applied. Thus, the age of this early hominid pushed back the age of the earliest hominids well beyond that which previously had been suggested.

It was later determined that the Leakeys' hominid did not represent an entirely new genus but was simply a very robust species of *Australopithecus*, and it was accordingly reclassified as *Australopithecus boisei*, and it is believed to have lived roughtly between 2 million and 1.2 million years ago.

Perhaps the most important impact of the Leakey team was the attention they drew to hominid research. While the Leakeys had become well known among their scientific peers in archaeology, prehistory, and paleontology, the public became equally familiar with their work after their discovery of *Zinjanthropus*. The discussion of this discovery—complete with color photographs and Louis Leakey's personal account in the September, 1960, issue of *National Geographic* magazine—played an important role in obtaining public support in the quest to document the human paleontological record. The support offered by the National Geographic Society led to the doubling of the excavation work being conducted at Olduvai. The increase in the recognition and support of paleontology had a dramatic impact on the scientific search for human fossil ancestors.

See also *Australopithecus*; Cro-Magnon Man; Gran Dolina Boy; Human Evolution; Langebaan Footprints; Lascaux Cave Paintings; Lucy; Neanderthals; Peking Man; Qafzeh Hominids.

FURTHER READING

Johanson, Donald C., and Maitland A. Edey. *Lucy: The Beginnings of Humankind*. New York: Simon & Schuster, 1981.
Johanson, Donald C., and James Shreeve. *Lucy's Child: The Discovery of a Human Ancestor*. New York: William Morrow, 1989.
Leakey, Louis S. B. *By the Evidence: Memoirs, 1932-1951*. New York: Harcourt Brace Jovanovich, 1974.
_____. "Finding the World's Earliest Man." *National Geographic* 118 (September, 1960): 420-435.
_____. "A New Fossil Skull from Olduvai." *Nature* 84 (August, 1959): 491-493.
Leakey, Mary D. *Disclosing the Past*. New York: Doubleday, 1984.
Leakey, Richard E., and Roger Lewin. *Origins*. New York: E. P. Dutton, 1977.
Morell, V. *Ancestral Passions: The Leakey Family and the Quest for Humankind's Beginnings*. New York: Simon & Schuster, 1995.
—*Turhon A. Murad, updated by Christina J. Moose*

Nobel Prize Science Laureates

Chemistry

1901	Jacobus H. van't Hoff
1902	Emil Fischer
1903	Svante Arrhenius
1904	Sir William Ramsay
1905	Adolf von Baeyer
1906	Henri Moissan
1907	Eduard Buchner
1908	Ernest Rutherford
1909	Wilhelm Ostwald
1910	Otto Wallach
1911	Marie Curie
1912	Victor Grignard, Paul Sabatier
1913	Alfred Werner
1914	Theodore W. Richards
1915	Richard Willstätter
1916	[Prize held in special fund]
1917	[Prize held in special fund]
1918	Fritz Haber
1919	[Prize held in special fund]
1920	Walther Nernst
1921	Frederick Soddy
1922	Francis W. Aston
1923	Fritz Pregl
1924	[Prize held in special fund]
1925	Richard Zsigmondy
1926	Theodor Svedberg
1927	Heinrich Wieland
1928	Adolf Windaus
1929	Arthur Harden, Hans von Euler-Chelpin
1930	Hans Fischer
1931	Carl Bosch, Friedrich Bergius
1932	Irving Langmuir
1933	[One-third to main fund, two-thirds to special fund]
1934	Harold C. Urey
1935	Frédéric Joliot, Irène Joliot-Curie
1936	Peter Debye

1937	Norman Haworth, Paul Karrer
1938	Richard Kuhn
1939	Adolf Butenandt, Leopold Ruzicka
1940	[One-third to main fund, two-thirds to special fund]
1941	[One-third to main fund, two-thirds to special fund]
1942	[One-third to main fund, two-thirds to special fund]
1943	George de Hevesy
1944	Otto Hahn
1945	Artturi Virtanen
1946	James B. Sumner, John H. Northrop, Wendell M. Stanley
1947	Sir Robert Robinson
1948	Arne Tiselius
1949	William F. Giauque
1950	Otto Diels, Kurt Alder
1951	Edwin M. McMillan, Glenn T. Seaborg
1952	Archer J. P. Martin, Richard L. M. Synge
1953	Hermann Staudinger
1954	Linus Pauling
1955	Vincent du Vigneaud
1956	Sir Cyril Hinshelwood, Nikolay Semenov
1957	Lord Todd
1958	Frederick Sanger
1959	Jaroslav Heyrovsky
1960	Willard F. Libby
1961	Melvin Calvin
1962	Max F. Perutz, John C. Kendrew
1963	Karl Ziegler, Giulio Natta
1964	Dorothy Crowfoot Hodgkin
1965	Robert B. Woodward
1966	Robert S. Mulliken
1967	Manfred Eigen, Ronald G. W. Norrish, George Porter
1968	Lars Onsager
1969	Derek Barton, Odd Hassel
1970	Luis Leloir
1971	Gerhard Herzberg
1972	Christian Anfinsen, Stanford Moore, William H. Stein
1973	Ernst Otto Fischer, Geoffrey Wilkinson
1974	Paul J. Flory
1975	John Cornforth, Vladimir Prelog
1976	William Lipscomb
1977	Ilya Prigogine

1978   Peter Mitchell
1979   Herbert C. Brown, Georg Wittig
1980   Paul Berg, Walter Gilbert, Frederick Sanger
1981   Kenichi Fukui, Roald Hoffmann
1982   Aaron Klug
1983   Henry Taube
1984   Bruce Merrifield
1985   Herbert A. Hauptman, Jerome Karle
1986   Dudley R. Herschbach, Yuan T. Lee, John C. Polanyi
1987   Donald J. Cram, Jean-Marie Lehn, Charles J. Pedersen
1988   Johann Deisenhofer, Robert Huber, Hartmut Michel
1989   Sidney Altman, Thomas R. Cech
1990   Elias James Corey
1991   Richard R. Ernst
1992   Rudolph A. Marcus
1993   Kary B. Mullis, Michael Smith
1994   George A. Olah
1995   Paul J. Crutzen, Mario J. Molina, F. Sherwood Rowland
1996   Robert F. Curl, Jr., Sir Harold Kroto, Richard E. Smalley
1997   Paul D. Boyer, John E. Walker, Jens C. Skou
1998   Walter Kohn, John Pople
1999   Ahmed Zewail
2000   Alan Heeger, Alan G. MacDiarmid, Hideki Shirakawa
2001   William S. Knowles, Ryoji Noyori, K. Barry Sharpless
2002   John B. Fenn, Koichi Tanaka, Kurt Wüthrich
2003   Peter Agre, Roderick MacKinnon
2004   Aaron Ciechanover, Avram Hershko, Irwin Rose
2005   Yves Chauvin, Robert H. Grubbs, Richard R. Schrock

Physics

1901   Wilhelm Conrad Röntgen
1902   Hendrik A. Lorentz, Pieter Zeeman
1903   Henri Becquerel, Pierre Curie, Marie Curie
1904   Lord Rayleigh
1905   Philipp Lenard
1906   J. J. Thomson
1907   Albert A. Michelson
1908   Gabriel Lippmann
1909   Guglielmo Marconi, Ferdinand Braun
1910   Johannes Diderik van der Waals

1911   Wilhelm Wien
1912   Gustaf Dalén
1913   Heike Kamerlingh Onnes
1914   Max von Laue
1915   William Bragg, Lawrence Bragg
1916   [Prize held in special fund]
1917   Charles Glover Barkla
1918   Max Planck
1919   Johannes Stark
1920   Charles Edouard Guillaume
1921   Albert Einstein
1922   Niels Bohr
1923   Robert A. Millikan
1924   Manne Siegbahn
1925   James Franck, Gustav Hertz
1926   Jean Baptiste Perrin
1927   Arthur H. Compton, C. T. R. Wilson
1928   Owen Willans Richardson
1929   Louis de Broglie
1930   Venkata Raman
1931   [Prize held in special fund]
1932   Werner Heisenberg
1933   Erwin Schrödinger, Paul A. M. Dirac
1934   [One-third to main fund, two-thirds to special fund]
1935   James Chadwick
1936   Victor F. Hess, Carl D. Anderson
1937   Clinton Davisson, George Paget Thomson
1938   Enrico Fermi
1939   Ernest Lawrence
1940   [One-third to main fund, two-thirds to special fund]
1941   [One-third to main fund, two-thirds to special fund]
1942   [One-third to main fund, two-thirds to special fund]
1943   Otto Stern
1944   Isidor Isaac Rabi
1945   Wolfgang Pauli
1946   Percy W. Bridgman
1947   Edward V. Appleton
1948   Patrick M. S. Blackett
1949   Hideki Yukawa
1950   Cecil Powell
1951   John Cockcroft, Ernest T. S. Walton

1952 Felix Bloch, E. M. Purcell
1953 Frits Zernike
1954 Max Born, Walther Bothe
1955 Willis E. Lamb, Polykarp Kusch
1956 William B. Shockley, John Bardeen, Walter H. Brattain
1957 Chen Ning Yang, Tsung-Dao Lee
1958 Pavel A. Cherenkov, Il'ja M. Frank, Igor Y. Tamm
1959 Emilio Segrè, Owen Chamberlain
1960 Donald A. Glaser
1961 Robert Hofstadter, Rudolf Mössbauer
1962 Lev Landau
1963 Eugene Wigner, Maria Goeppert-Mayer, J. Hans D. Jensen
1964 Charles H. Townes, Nicolay G. Basov, Aleksandr M. Prokhorov
1965 Sin-Itiro Tomonaga, Julian Schwinger, Richard P. Feynman
1966 Alfred Kastler
1967 Hans Bethe
1968 Luis Alvarez
1969 Murray Gell-Mann
1970 Hannes Alfvén, Louis Néel
1971 Dennis Gabor
1972 John Bardeen, Leon N. Cooper, Robert Schrieffer
1973 Leo Esaki, Ivar Giaever, Brian D. Josephson
1974 Martin Ryle, Antony Hewish
1975 Aage N. Bohr, Ben R. Mottelson, James Rainwater
1976 Burton Richter, Samuel C. C. Ting
1977 Philip W. Anderson, Sir Nevill F. Mott, John H. van Vleck
1978 Pyotr Kapitsa, Arno Penzias, Robert Woodrow Wilson
1979 Sheldon Glashow, Abdus Salam, Steven Weinberg
1980 James Cronin, Val Fitch
1981 Nicolaas Bloembergen, Arthur L. Schawlow, Kai M. Siegbahn
1982 Kenneth G. Wilson
1983 Subramanyan Chandrasekhar, William A. Fowler
1984 Carlo Rubbia, Simon van der Meer
1985 Klaus von Klitzing
1986 Ernst Ruska, Gerd Binnig, Heinrich Rohrer
1987 J. Georg Bednorz, K. Alex Müller
1988 Leon M. Lederman, Melvin Schwartz, Jack Steinberger
1989 Norman F. Ramsey, Hans G. Dehmelt, Wolfgang Paul
1990 Jerome I. Friedman, Henry W. Kendall, Richard E. Taylor
1991 Pierre-Gilles de Gennes
1992 Georges Charpak

1993   Russell A. Hulse, Joseph H. Taylor, Jr.
1994   Bertram N. Brockhouse, Clifford G. Shull
1995   Martin L. Perl, Frederick Reines
1996   David M. Lee, Douglas D. Osheroff, Robert C. Richardson
1997   Steven Chu, Claude Cohen-Tannoudji, William D. Phillips
1998   Robert B. Laughlin, Horst L. Störmer, Daniel C. Tsui
1999   Gerardus 't Hooft, Martinus J. G. Veltman
2000   Zhores I. Alferov, Herbert Kroemer, Jack S. Kilby
2001   Eric A. Cornell, Wolfgang Ketterle, Carl E. Wieman
2002   Raymond Davis, Jr., Masatoshi Koshiba, Riccardo Giacconi
2003   Alexei A. Abrikosov, Vitaly L. Ginzburg, Anthony J. Leggett
2004   David J. Gross, H. David Politzer, Frank Wilczek
2005   Roy J. Glauber, John L. Hall, Theodor W. Hänsch

PHYSIOLOGY OR MEDICINE

1901   Emil von Behring
1902   Ronald Ross
1903   Niels Ryberg Finsen
1904   Ivan Pavlov
1905   Robert Koch
1906   Camillo Golgi, Santiago Ramón y Cajal
1907   Alphonse Laveran
1908   Ilya Mechnikov, Paul Ehrlich
1909   Theodor Kocher
1910   Albrecht Kossel
1911   Allvar Gullstrand
1912   Alexis Carrel
1913   Charles Richet
1914   Robert Bárány
1915   [Prize held in special fund]
1916   [Prize held in special fund]
1917   [Prize held in special fund]
1918   [Prize held in special fund]
1919   Jules Bordet
1920   August Krogh
1921   [Prize held in special fund]
1922   Archibald V. Hill, Otto Meyerhof
1923   Frederick G. Banting, John Macleod
1924   Willem Einthoven
1925   [Prize held in special fund]

1926   Johannes Fibiger
1927   Julius Wagner-Jauregg
1928   Charles Nicolle
1929   Christiaan Eijkman, Sir Frederick Hopkins
1930   Karl Landsteiner
1931   Otto Warburg
1932   Sir Charles Sherrington, Edgar Adrian
1933   Thomas H. Morgan
1934   George H. Whipple, George R. Minot, William P. Murphy
1935   Hans Spemann
1936   Sir Henry Dale, Otto Loewi
1937   Albert Szent-Györgyi
1938   Corneille Heymans
1939   Gerhard Domagk
1940   [One-third to main fund, two-thirds to special fund]
1941   [One-third to main fund, two-thirds to special fund]
1942   [One-third to main fund, two-thirds to special fund]
1943   Henrik Dam, Edward A. Doisy
1944   Joseph Erlanger, Herbert S. Gasser
1945   Sir Alexander Fleming, Ernst B. Chain, Sir Howard Florey
1946   Hermann J. Muller
1947   Carl Cori, Gerty Cori, Bernardo Houssay
1948   Paul Müller
1949   Walter Hess, Egas Moniz
1950   Edward C. Kendall, Tadeus Reichstein, Philip S. Hench
1951   Max Theiler
1952   Selman A. Waksman
1953   Hans Krebs, Fritz Lipmann
1954   John F. Enders, Thomas H. Weller, Frederick C. Robbins
1955   Hugo Theorell
1956   André F. Cournand, Werner Forssmann, Dickinson W. Richards
1957   Daniel Bovet
1958   George Beadle, Edward Tatum, Joshua Lederberg
1959   Severo Ochoa, Arthur Kornberg
1960   Sir Frank Macfarlane Burnet, Peter Medawar
1961   Georg von Békésy
1962   Francis Crick, James Watson, Maurice Wilkins
1963   Sir John Eccles, Alan L. Hodgkin, Andrew F. Huxley
1964   Konrad Bloch, Feodor Lynen
1965   François Jacob, André Lwoff, Jacques Monod
1966   Peyton Rous, Charles B. Huggins

1967	Ragnar Granit, Haldan K. Hartline, George Wald
1968	Robert W. Holley, H. Gobind Khorana, Marshall W. Nirenberg
1969	Max Delbrück, Alfred D. Hershey, Salvador E. Luria
1970	Sir Bernard Katz, Ulf von Euler, Julius Axelrod
1971	Earl W. Sutherland, Jr.
1972	Gerald M. Edelman, Rodney R. Porter
1973	Karl von Frisch, Konrad Lorenz, Nikolaas Tinbergen
1974	Albert Claude, Christian de Duve, George E. Palade
1975	David Baltimore, Renato Dulbecco, Howard M. Temin
1976	Baruch S. Blumberg, D. Carleton Gajdusek
1977	Roger Guillemin, Andrew V. Schally, Rosalyn Yalow
1978	Werner Arber, Daniel Nathans, Hamilton O. Smith
1979	Allan M. Cormack, Godfrey N. Hounsfield
1980	Baruj Benacerraf, Jean Dausset, George D. Snell
1981	Roger W. Sperry, David H. Hubel, Torsten N. Wiesel
1982	Sune K. Bergström, Bengt I. Samuelsson, John R. Vane
1983	Barbara McClintock
1984	Niels K. Jerne, Georges J. F. Köhler, César Milstein
1985	Michael S. Brown, Joseph L. Goldstein
1986	Stanley Cohen, Rita Levi-Montalcini
1987	Susumu Tonegawa
1988	Sir James W. Black, Gertrude B. Elion, George H. Hitchings
1989	J. Michael Bishop, Harold E. Varmus
1990	Joseph E. Murray, E. Donnall Thomas
1991	Erwin Neher, Bert Sakmann
1992	Edmond H. Fischer, Edwin G. Krebs
1993	Richard J. Roberts, Phillip A. Sharp
1994	Alfred G. Gilman, Martin Rodbell
1995	Edward B. Lewis, Christiane Nüsslein-Volhard, Eric F. Wieschaus
1996	Peter C. Doherty, Rolf M. Zinkernagel
1997	Stanley B. Prusiner
1998	Robert F. Furchgott, Louis J. Ignarro, Ferid Murad
1999	Günter Blobel
2000	Arvid Carlsson, Paul Greengard, Eric R. Kandel
2001	Leland H. Hartwell, Tim Hunt, Sir Paul Nurse
2002	Sydney Brenner, H. Robert Horvitz, John E. Sulston
2003	Paul C. Lauterbur, Sir Peter Mansfield
2004	Richard Axel, Linda B. Buck
2005	Barry J. Marshall, J. Robin Warren

Time Line

The more than 750 events below represent milestones in the major sciences, theoretical and applied, from ancient times to 2005.

585 B.C.E.	Thales of Miletus, a Greek philosopher, predicts a solar eclipse. About the same time he theorizes that water is the fundamental element for all substances.
c. 550 B.C.E.	Construction of trireme changes naval warfare.
c. 550 B.C.E.	Greek philosopher and astronomer Anaximander proposes a theory of biological evolution.
c. 530 B.C.E.	Greek mathematician and philosopher Pythagoras invents the Pythagorean theorem. He also argues that the Earth is a sphere and that the Sun, stars, and planets revolve around it.
c. 500 B.C.E.	Chinese physicians begin the practice of acupuncture.
500 B.C.E.	Some Greek city-states take care of sick people in hospitals called *aesculapia* (named after Aesclepius, the god of medicine).
c. 500 B.C.E.	The Greek physician and scientist Alcmaeon of Croton makes the first known dissections of dead human bodies.
5th cent. B.C.E.	Greek philosopher Anaxagoras writes *On Nature*, arguing that mind exists and that matter is composed of an infinite number of atomic elements.
c. 430 B.C.E.	Death of Greek philosopher Empedocles, who held that all matter is made of four elements: water, fire, air, and earth.
c. 400 B.C.E.	Greek philosopher Philolaus is the first known person to argue that Earth orbits around the Sun.
c. 370 B.C.E.	Death of Greek physician Hippocrates, author of many books with detailed case histories and proposed physical explanations for diseases. The Hippocratic Oath, which appears later, represents his principles.
c. 325 B.C.E.	Greek physician Praxagoras of Cos discovers the value of measuring the pulse when diagnosing diseases.
c. 323 B.C.E.	Aristotle theorizes about the nature of species, reproduction, and hybrids.

c. 320 B.C.E.	Theophrastus initiates the study of botany.
312 B.C.E.	First Roman aqueduct is built.
300 B.C.E.	Babylonian mathematicians develop a symbol for zero.
c. 300 B.C.E.	Greek mathematician Euclid of Egypt writes *Elements*, which includes a summary of plane and solid geometry.
c. 300 B.C.E.	*The Yellow Emperor's Classic of Internal Medicine*, a compilation attributed to Chinese emperor Huangdi, contains references to the function of the heart and the circulation of the blood.
Early 3d cent. B.C.E.	Greek astronomer Aristarchus of Samos writes *On the Size and Distance of the Sun and the Moon*, arguing that the Earth revolves around the Sun.
c. 250 B.C.E.	Greek scientist Archimedes discovers the law of specific gravity, later known as Archimedes' principle.
From 240 B.C.E.	Romans learn to use the arch in building.
240 B.C.E.	Chinese astronomers make the first known observation of Halley's comet.
240 B.C.E.	Eratosthenes of Cyrene, librarian of Alexandria, Egypt, correctly calculates the circumference of the Earth at about 25,000 miles.
221 B.C.E.-220 C.E.	Advances are made in Chinese agricultural technology.
200 B.C.E.	The Greeks invent the astrolabe to determine the positions of the stars.
165 B.C.E.	The Chinese make the first known observations of sunspots.
150 B.C.E.	Greek astronomer Hipparchus of Nicaea calculates that the Moon is about 240,000 miles from the Earth.
100 B.C.E.	The Romans begin to use water power to mill flour.
c. 100 B.C.E.	Greek philosopher Poseidonius shows correlation between tides and the lunar cycle.
46 B.C.E.	Establishment of the Julian calendar.
7 B.C.E.	Greek philosopher Strabo summarizes geographical knowledge in his *Geography*.
c. 62 C.E.	Hero of Alexander invents a simple steam engine, which is never found to have a practical use.

77 C.E.	Roman natural philosopher Pliny the Elder publishes *Natural History*, which will serve as a standard scientific handbook until the Renaissance.
2d cent. C.E.	The Daoist religious leader Zhang Daoling composes a guide of charms and incantations that presumably cure diseases.
c. 105 C.E.	Chinese inventor Cai Lun makes paper out of wood, rags, or other materials containing cellulose.
c. 150 C.E.	Alexandrian scientist Ptolemy argues that all heavenly bodies revolve around a fixed Earth in the *Almagest*.
c. 157-201 C.E.	Greek physician and anatomist Galen proves that the arteries carry blood but incorrectly explains how the blood passes through the heart.
c. 250 C.E.	The Maya in Mexico and Central America are beginning scientific and technological advances that will continue for about six hundred years.
c. 350 C.E.	The Chinese invent an early form of printing.
369 C.E.	Saint Basil erects a hospital at Caesarea.
c. 400 C.E.	The Chinese invent the wheelbarrow.
563	Silk worms are smuggled to the Byzantine Empire.
595-665	Invention of decimals and negative numbers.
7th-early 8th cent.	Maya build astronomical observatory at Palenque.
7th-8th cent.	Papermaking spreads to Korea, Japan, and Central Asia.
c. 700	The bow and arrow spread into North America.
c. 700-1000	The heavy plow increases agricultural yields.
Mid-9th cent.	Invention of firearms using gunpowder.
c. 1045	In China, Bi Sheng develops movable earthenware type.
c. 1200	Development of scientific cattle-breeding techniques.
1275	Invention of the first mechanical clock.
1328	Thomas Bradwardine's *Treatise on Proportions* begins a period of intense investigation at Merton College, Oxford, into what would later be called the laws of physics.
c. 1450-1456	Gutenberg pioneers the printing press, culminating in the publication of Gutenberg's Mazarin Bible.

1462 Regiomontanus (Johann Müller) completes the *Epitome* of Ptolemy's *Almagest*. His work forms the basis of trigonometry in Western Europe as handed down from the Arab world, and his astronomical observations and charts will be used by Christopher Columbus.

1474 Great Wall of China is built.

c. 1478-1519 Leonardo da Vinci compiles his notebooks.

1490's Aldus Manutius founds the Aldine Press.

Beginning 1490 Development of the camera obscura.

1490-1492 Martin Behaim builds the first world globe.

16th cent. Evolution of the galleon.

16th cent. Proliferation of firearms.

c. 1510 Invention of the watch.

1517 Fracastoro develops his theory of fossils.

1530's-1540's Paracelsus presents his theory of disease.

1543 Copernicus publishes *De revolutionibus*, articulating his heliocentric view of the universe.

1543 Vesalius publishes *On the Fabric of the Human Body*, which will be used for human anatomical studies for generations.

1546 Fracastoro discovers that contagion spreads disease.

1550's Tartaglia publishes *The New Science*.

1553 Michael Servetus describes the circulatory system.

c. 1560's Invention of the "lead" (graphite) pencil.

1569 Mercator publishes his world map.

1572-1574 Tycho Brahe observes a supernova and conducts astronomical observations and measurements on which Johannes Kepler will base much of his work.

1580's-1590's Galileo conducts his early experiments in motion and falling bodies.

1582 Gregory XIII reforms the Western calendar.

1600 William Gilbert publishes *De Magnete*, pioneering the study of magnetism and Earth's magnetic field.

17th cent. England undergoes an agricultural revolution.

17th cent.	Rise of the "gunpowder empires" with a new type of warfare and geopolitical relations based on advances in firearms and gunpowder-based weaponry.
1601-1672	Rise of European scientific societies, which join great mathematical and scientific minds with official approval and institutionalized support.
Sept., 1608	Hans Lippershey invents a simple telescope, credited as the first.
1609-1619	Johannes Kepler develops his laws of planetary motion.
1610	Galileo confirms the heliocentric model of the solar system.
1612	Sanctorius (Santorio) invents the clinical thermometer.
1615-1696	Sir Isaac Newton and Gottfried Wilhelm Leibniz independently invent the calculus.
1620	Sir Francis Bacon publishes *Novum Organum*, in which he advocates an inductive, empirical scientific method.
1623-1674	Appearance of the earliest calculators.
1629	In Persia (Iran), the Ṣafavid Dynasty flourishes under ʿAbbās the Great.
1632	Galileo publishes *Dialogue Concerning the Two Chief World Systems, Ptolemaic and Copernican*.
1637	René Descartes publishes *Discourse on Method*, articulating the Cartesian scientific method.
1638	The printing press arrives in North America.
1643	Evangelista Torricelli measures atmospheric pressure.
1651	William Harvey suggests that all living things must originate in an egg.
1655-1663	Francesco Grimaldi discovers the principle of light diffraction.
Feb., 1656	Christiaan Huygens identifies the rings of Saturn.
1660's-1700	Antoni van Leeuwenhoek and others conduct the first observations using microscopes.

1660-1692   The "father of modern chemistry," Robert Boyle, discovers the inverse relationship between the pressure and volume of a gas and uses a corpuscular (atomic) theory of matter to explain his experimental results.

1664   Thomas Willis identifies the basal ganglia.

1665   Gian Domenico Cassini discovers Jupiter's Great Red Spot.

1669   Nicholas Steno, the "father of stratigraphy," presents his theories of fossils and dynamic geology.

c. 1670   First widespread smallpox inoculations using a method imported from the Ottoman Empire, variolation.

Late Dec., 1671   Sir Isaac Newton builds the first reflecting telescope.

1673   Christiaan Huygens explains the pendulum.

1676   Thomas Sydenham advocates clinical observation.

Dec. 7, 1676   Ole Rømer calculates the speed of light.

1677   Antoni van Leeuwenhoek describes sperm and eggs and collects evidence that helps disprove the theory of spontaneous generation.

1686   Edmond Halley develops the first weather map.

Summer, 1687   Sir Isaac Newton publishes his *Principia*, the most important scientific treatise of the century, in which he presents his theory of universal gravitation.

1691-1694   German botanist Rudolph Jacob Camerarius establishes the existence of sex in plants.

July 25, 1698   Thomas Savery patents the first successful steam engine.

1701   Jethro Tull invents the seed drill.

1704   Sir Isaac Newton publishes *Opticks*.

1705   Edmond Halley predicts the return of his comet.

1705-1712   Thomas Newcomen develops the steam engine.

1709   Darby invents coke-smelting of iron ore.

1714   Mill patents the typewriter.

1714   Daniel Fahrenheit develops the mercury thermometer.

1714-1735   The quest for a means of determining longitude at sea leads John Harrison to develop his chronometer.

1718   Publication of Daniel Bernoulli's *Calculus of Variations*.

1718   Geoffroy issues the *Table of Affinites*.

1722   René-Antoine Réaumur discovers carbon's role in hardening steel.

1722-1733   Abraham de Moivre describes the bell-shaped curve.

1723   Stahl postulates phlogiston as the basis for combustion.

1724   The St. Petersburg Academy of Sciences is established in Russia.

1725   John Flamsteed, Britain's first astronomer royal, issues the first comprehensive star catalog, *Historia Coelistis Britannica*.

1729   Stephen Gray discovers the principles of electric conduction.

1733   Charles Du Fay describes positive and negative electric charge.

1733   John Kay invents the flying shuttle.

1735   George Hadley describes atmospheric circulation.

1735   Carl Linnaeus creates the binomial system of classification of plants and animals.

1735-1743   Charles La Condamine measures a meridional arc at the equator and explores the Amazon River basin.

1738   Daniel Bernoulli proposes the kinetic theory of gases.

1742   Anders Celsius proposes an international fixed temperature scale.

1743-1744   Jean le Rond d'Alembert develops his axioms of motion.

1745   Invention of the Leyden jar.

1745   Mikhail Lomonosov issues the first catalog of minerals.

1746   John Roebuck develops the lead-chamber process.

1747   Andreas Marggraf extracts sugar from beets.

1748   James Bradley discovers the nutation of Earth's axis.

1748   Jean-Antoine Nollet discovers osmosis.

1748   Maria Agnesi publishes *Analytical Institutions,* a two-volume textbook on the calculus that offers a complete synthesis of the mathematical methods developed in the scientific revolution.

1748   Leonhard Euler develops integral calculus.

1749-1789   Georges Leclerc (comte de Buffon) publishes his *Natural History,* the first comprehensive examination of the natural world.

1751   Pierre Louis de Maupertuis postulates "hereditary particles" as the basis for inherited traits.

1752   Benjamin Franklin demonstrates the electrical nature of lightning.

1752   Johann Tobias Mayer's lunar tables enable mariners to determine longitude at sea.

1753   James Lind identifies citrus fruit as a preventive for scurvy.

1755   Joseph Black identifies carbon dioxide.

1757   Alexander Monro distinguishes between lymphatic and blood systems.

1759   Franz Aepinus publishes *Theory of Electricity and Magnetism.*

1760's   Robert Bakewell begins selective livestock breeding.

1764   James Hargreaves invents the spinning jenny, which dramatically increases the output of the textile industry.

1764   The Reverend Thomas Bayes issues his "Essay Towards Solving a Problem in the Doctrine of Chances," on inverse probability.

1765-1769   James Watt develops his steam engine.

1766   Albrecht von Haller publishes *Elements of Human Physiology.*

1767-1768   Lazzaro Spallanzani refutes the theory of spontaneous generation.

1768-1771   Richard Arkwright develops the water frame.

1769-1770   Nicolas Cugnot builds a steam-powered road carriage.

1771   Discovery of picric acid and its explosive properties.

1772-1789   Antoine-Laurent Lavoisier devises the modern system of chemical nomenclature.

c. 1773 Sir William Herschel builds his reflecting telescope.

1774 Joseph Priestley discovers oxygen.

1776 First test of a submarine in warfare.

1777 Jan Ingenhousz discovers photosynthesis.

1779 Compton invents the spinning mule.

1783 Henry Cort improves iron processing.

1783 Nicolas Leblanc develops a process for producing soda from common salt.

1784 Adrien-Marie Legendre introduces polynomials.

1784-1785 Henry Cavendish discovers the composition of water.

1784-1788 Andrew Meikle invents the drum thresher.

1785 Edmund Cartwright invents the steam-powered loom.

1785-1788 James Hutton proposes the uniformitarian theory of the history of Earth and geologic change.

1789 Invention of the guillotine.

1790 Samuel Slater invents a cotton-spinning mill and will become known as the father of the American cotton industry.

1793 Eli Whitney invents the cotton gin.

1795 Invention of the flax spinner.

1796 Pierre-Simon Laplace articulates his nebular hypothesis.

1796-1798 Edward Jenner develops smallpox vaccination.

1798 Thomas Robert Malthus publishes *An Essay on the Principle of Population*.

1799 Discovery of the earliest anesthetics.

1799 Joseph Louis Proust establishes law of definite proportions, thus effectively distinguishing between chemical elements and chemical compounds.

1800 Alessandro Volta invents the battery.

1801 Astronomers make the first discovery of an asteroid, Ceres.

1803-1807 John Dalton formulates the atomic theory of matter.

1804 Nicolas de Saussure publishes *Chemical Research in Vegetation*.

1804 First successful steam locomotive runs in Wales.

c. 1805   William H. Wollaston develops principles of modern metallurgy and later discovers the dark lines in the solar spectrum.

1809   Sir Humphry Davy invents the arc lamp.

1809   Jean-Baptiste Lamarck publishes *Zoological Philosophy,* in which he sets forth his law of acquired characteristics.

1814   Joseph Fraunhofer invents the spectroscope.

1816   René Laennec invents the stethoscope.

1818-1843   Triangulation Survey of India.

1820's   André Ampère reveals magnetism's relationship to electricity.

1823   William Buckland conducts early studies of dinosaurs.

1829   Louis Braille invents printing for the blind.

1830   Sir Charles Lyell publishes *Principles of Geology.*

1831   Michael Faraday converts magnetic force into electricity.

1831   Cyrus Hall McCormick invents the reaper.

1835   Charles Babbage invents a mechanical calculator.

1836   Samuel Colt patents the revolver.

1838   G. J. Mulder precipitates a fibrous material from cells, which he calls "protein."

1838-1839   Matthias Schleiden and Theodor Schwann's cell theory becomes the foundation of modern biology.

1839   Louis Daguerre and Joseph Niepce invent daguerreotype photography.

1840   Justus von Liebig invents artificial fertilizers.

1844   Charles Goodyear patents vulcanized rubber.

1846   Elias Howe patents his sewing machine.

1846   First demonstration of surgical anesthesia by ether inhalation.

1847   Ignaz Philipp Semmelweis recognizes that puerperal fever is spreading by transmission from doctors, nurses, and medical students within his hospital and advances antiseptic practices by insisting on hand washing.

1847   George Boole publishes *Mathematical Analysis of Logic*, establishing the field of mathematical logic.

1850   Rudolf Clausius formulates second law of thermodynamics.

1850   The giant moa becomes extinct.

1850   Theodore Schwann, Matthias Schleiden, and Rudolf Virchow recognize that tissues are made up of cells.

1855   Florence Nightingale reforms nursing in the Crimea.

1855   Alfred Russel Wallace publishes *On the Law Which Has Regulated the Introduction of New Species*, in which he develops the theory of natural selection around the same time as Charles Darwin.

1855-1859   Sir Henry Bessemer develops new methods for processing steel.

1856   A Neanderthal skull is found near Düsseldorf.

1856   Louis Pasteur begins research into fermentation, later developing his "pasteurization" process.

1858   Étienne Lenoir invents the internal combustion engine.

1859   Charles Darwin publishes *On the Origin of Species by Means of Natural Selection*, in which he sets forth his theory of natural selection, the mechanism of evolution.

1861   The oldest bird fossil, *Archaeopteryx*, is discovered at Solnhofen.

1864-1867   Joseph Lister promotes antiseptic surgery.

1866   Alfred Nobel invents dynamite.

1866   Ernst Haeckel develops the hypothesis that hereditary information is transmitted by the cell nucleus.

1866   Gregor Mendel, an Austrian monk, publishes a paper introducing his ideas of the mechanisms of heredity, including dominant and recessive traits.

1868   The bones of a Cro-Magnon skeleton, thought to be the earliest modern human being, are discovered in France.

1868   Christopher Latham Sholes patents a practical typewriter.

1869   George Westinghouse patents air brakes.

1869 Friedrich Miescher isolates "nuclein" from the nuclei of white blood cells, which is later found to be the nucleic acids DNA and RNA.

1869 Dmitry Mendeleyev develops the periodic table of elements.

1871 Darwin Publishes *The Descent of Man and Selection in Relation to Sex.*

1873 Jean-Martin Charcot publishes *Leçons sur les maladies du systeme nerveux.*

1875 Oskar Hertwig demonstrates the fertilization of an ovum in a sea urchin, establishing the principle of sexual reproduction: the union of egg and sperm cells.

1876 Alexander Graham Bell demonstrates the telephone.

1876 Nikolaus Otto invents a practical internal combustion engine.

1877 Thomas Alva Edison patents the cylinder phonograph.

1878 Eadweard Muybridge uses photography to study animal movement.

1879 Thomas Alva Edison demonstrates the incandescent lamp, an early form of the lightbulb.

1880 Walter Fleming first describes mitosis.

1882 The first birth control clinic is established in Amsterdam.

1882-1884 Robert Koch isolates microorganisms that cause tuberculosis and cholera.

1883 Wilhelm Roux theorizes that mitosis must result in equal sharing of all chromosomal particles.

1883 Francis Galton founds the field of eugenics.

1884 Hiram Stevens Maxim improves the machine gun.

1885 Carl Benz develops the first practical automobile.

1887 Hannibal Williston Goodwin develops celluloid film.

1887-1890 Theodor Boveri notes that chromosomes are preserved through cell division and that sperm and egg contribute equal numbers of chromosomes.

1888 John Boyd Dunlop patents the pneumatic tire.

1888-1906	Santiago Ramón y Cajal establishes the neuron as the functional unit of the nervous system.
1890-1901	Emil von Behring discovers the diphtheria antitoxin.
1893	Rudolf Diesel patents the diesel engine.
1895	Wilhelm Röntgen discovers X rays.
1896	Guglielmo Marconi patents the telegraph.
1897	Felix Hoffman invents aspirin.
1897	Sir Ronald Ross discovers the malaria bacillus.
1897-1901	John Jacob Abel and Jokichi Takamine independently isolate adrenaline.
July, 1897-July, 1904	Vilhelm Bjerknes publishes the first weather forecasting computational hydrodynamics.
1898	Martinus Beijerinck discovers viruses.
1898-1902	Teisserenc de Bort discovers the stratosphere and the troposphere.
Sept., 1898-July, 1900	David Hilbert develops a model for Euclidean geometry in arithmetic.
1899-1902	Henri-Léon Lebesgue develops new integration theory.
Early 1900's	Willem Einthoven develops the forerunner of the electrocardiogram.
1900	Sir Frederick Hopkins discovers tryptophan, an essential amino acid.
1900	Emil Wiechert invents the inverted pendulum seismography.
Mar.-June, 1900	Hugo de Vries and associates discover Gregor Johann Mendel's ignored studies of inheritance.
Mar. 23, 1900	Arthur Evans discovers the Minoan civilization on Crete.
1900-1901	Karl Landsteiner discovers human blood groups.
June, 1900-Feb., 1901	Walter Reed establishes that yellow fever is transmitted by mosquitoes.
July 2, 1900	Ferdinand von Zeppelin constructs the first dirigible that flies.
Dec. 14, 1900	Max Planck announces his quantum theory.
1901	Peter Cooper Hewitt invents the mercury vapor lamp.

1901   Julius Elster and Hans Friedrich Geitel demonstrate radioactivity in rocks, springs, and air.

1901   The first synthetic vat dye, indanthrene blue, is synthesized.

1901   Gerrit Grijns proposes that beriberi is caused by a nutritional deficiency.

1901   Ilya Ivanov develops artificial insemination.

Dec. 12, 1901   Guglielmo Marconi receives the first transatlantic telegraphic radio transmission.

1901-1904   Frederic Stanley Kipping discovers silicones.

1902   Eldridge R. Johnson perfects the process to mass-produce disc recordings.

1902   Beppo Levi recognizes the axiom of choice in set theory.

1902   Clarence McClung plays a role in the discovery of the sex chromosome.

1902   Walter S. Sutton states that chromosomes are paired and could be carriers of hereditary traits.

1902   Alexis Carrel develops a technique for rejoining severed blood vessels.

1902   Richard Zsigmondy invents the ultramicroscope.

1902   Arthur Edwin Kennelly and Oliver Heaviside propose the existence of the ionosphere.

1902-1903   Ivan Pavlov develops the concept of reinforcement.

Jan., 1902   The French expedition at Susa discovers the Hammurabi code.

Apr.-June, 1902   William Maddock Bayliss and Ernest Henry Starling discover secretin and establish the role of hormones.

June 16, 1902   Bertrand Russell discovers the "Great Paradox" concerning the set of all sets.

1903   Konstantin Tsiolkovsky proposes that liquid oxygen be used for space travel.

1903-1904   George Ellery Hale establishes Mount Wilson Observatory.

Sept. 10, 1903   Antoine-Henri Becquerel wins the Nobel Prize for the discovery of natural radioactivity.

Dec. 17, 1903   The Wright brothers launch the first successful airplane.

1904-1905	William Crawford gorgas develops effective methods for controlling mosquitoes.
1904-1907	L. E. J. Brouwer develops intuitionist foundations of mathematics.
1904-1908	Ernst Zermelo undertakes the first comprehensive axiomatization of set theory.
1904-1912	Jacques Edwin Brandenberger invents cellophane.
1904	Julius Elster and Hans Friedrich Geitel devise the first practical photoelectric cell.
1904	Johannes Franz Hartmann discovers the first evidence of interstellar matter.
1904	Jacobus Cornelis Kapteyn discovers two star streams in the galaxy.
Apr.-May, 1904	Sir Charles Scott Sherrington delivers *The Integrative Action of the Nervous System.*
Summer, 1904	Construction begins on the Panama Canal.
Nov. 16, 1904	Sir John Ambrose Fleming files a patent for the first vacuum tube.
1905	George Washington Crile performs the first direct blood transfusion.
1905	Albert Einstein develops his theory of the photoelectric effect.
1905-1907	Leo Hendrik Baekeland invents Bakelite.
1905-1907	Bertram Boltwood uses radioactivity to obtain the age of rocks.
1905	Ejnar Hertzsprung notes the relationship between color and luminosity of stars.
1905	Punnett's *Mendelism* presents his diagrams for showing how hereditary traits are passed from one generation to the next.
Aug., 1905	Percival Lowell predicts the existence of Pluto.
1906	Frederick Gardner Cottrell invents the electronstat precipitation process.
1906	Sir Frederick Hopkins suggests that food contains vitamins essential to life.
1906	Hermann Anschütz-Kaempfe installs a gyrocompass onto a German battleship.
1906	Charles Glover Barkla discovers the characteristic X rays of the elements.

1906	Maurice Fréchet introduces the concept of abstract space.
1906	Andrey Markov discovers the theory of linked probabilities.
1906-1910	Richard D. Oldham and Andrija Mohorovičić determine the structure of the Earth's interior.
1906-1913	Richard Willstätter discovers the composition of chlorophyll.
Aug. 4, 1906	The first German U-boat submarine is launched.
Dec., 1906	J. J. Thomson wins the Nobel Prize for the discovery of the electron.
Dec. 24, 1906	Reginald Aubrey Fessenden perfects radio by transmitting music and voice.
1907	Louis and Auguste Lumière develop color photography.
1907	John Scott Haldane develops stage decompression for deep-sea divers.
1907	Ejnar Hertzsprung describes giant and dwarf stellar divisions.
Spring, 1907	Ross Granville Harrison observes the development of nerve fibers in the laboratory.
1908	Fritz Haber develops a process for extracting nitrogen from the air.
1908	Hardy and Weinberg present a model of population genetics.
1908	Howard Hughes, Sr., revolutionizes oil-well drilling.
1908	Charles Proteus Steinmetz warns of pollution in *The Future of Electricity*.
1908-1915	Thomas Hunt Morgan develops the gene-chromosome theory.
Feb. 11, 1908	Hans Geiger and Ernest Rutherford develop the Geiger counter.
June 26, 1908	George Ellery Hale discovers strong magnetic fields in sunspots.
Nov.-Dec., 1908	Paul Ehrlich and Élie Metchnikoff conduct pioneering research in immunology.
Dec., 1908	Marcellin Boule reconstructs the first Neanderthal skeleton.

1909	Wilhelm Johannsen coins the terms "gene," "genotype," and "phenotype."
1909	The study of mathematical fields by Ernst Steinitz inaugurates modern abstract algebra.
Jan.-Aug., 1909	Robert Andrews Millikan conducts his oil-drop experiment.
July 25, 1909	Louis Blériot makes the first airplane flight across the English channel.
1910	The electric washing machine is introduced.
1910	Peyton Rous discovers that some cancers are caused by viruses.
1910	Bertrand Russell and Alfred North Whitehead's *Principia Mathematica* develops the logistic movement in mathematics.
1910	J. J. Thomson confirms the possibility of isotopes.
Apr., 1910	Paul Ehrlich introduces Salvarsan as a cure for syphilis.
1911	Franz Boas publishes *The Mind of Primitive Man*.
July 24, 1911	Hiram Bingham discovers an Inca city in the Peruvian jungle.
Fall, 1911	Alfred H. Sturtevant produces the first chromosome map.
1912	Henrietta Swan Leavitt's study of variable stars unlocks galactic distances.
1912	Vesto Slipher obtains the spectrum of a distant galaxy.
1912-1913	Niels Bohr writes a trilogy on atomic and molecular structure.
1912-1914	John Jacob Abel develops the first artificial kidney.
1912-1915	X-ray crystallography is developed by William Henry and Lawrence Bragg.
Jan., 1912	Alfred Lothar Wegener proposes the theory of continental drift.
Mar. 7, 1912	Ernest Rutherford presents his theory of the atom.
Aug. 7 and 12, 1912	Victor Franz Hess discovers cosmic rays through high-altitude ionizations.
1913	Thomas Alva Edison introduces the kinetophone to show the first talking pictures.

1913	Henry Ford produces automobiles on a moving assembly line.
1913	Ejnar Hertzsprung uses cepheid variables to calculate the distances to stars.
1913	Geothermal power is produced for the first time.
1913	Beno Gutenberg discovers the Earth's mantle-outer core boundary.
1913	Albert Salomon develops mammography.
1913	Béla Schick introduces the Schick test for diphtheria.
Jan., 1913	William Merriam Burton introduces thermal cracking for refining petroleum.
Jan. 17, 1913	Charles Fabry quantifies ozone in the upper atmosphere.
Dec., 1913	Henry Norris Russell announces his theory of stellar evolution.
1914	Ernest Rutherford discovers the proton.
Apr., 1915	The first transcontinental telephone call is made.
May, 1915	The Fokker aircraft are the first airplanes equipped with machine guns.
May 20, 1915	Corning Glass Works trademarks pyrex and offers pyrex cookware for commercial sale.
Sept., 1915-Feb., 1916	Jay McLean discovers the natural anticoagulant heparin.
Oct., 1915	Transatlantic radiotelephony is first demonstrated.
Oct., 1915-Mar., 1917	Paul Langevin develops active sonar for submarine detection and fathometry.
Nov. 25, 1915	Albert Einstein completes his theory of general relativity.
1916	Karl Schwarzschild develops a solution to the equations of general relativity.
1917	Clarence Birdseye develops freezing as a way of preserving foods.
1917	Insecticide use intensifies when arsenic proves effective against the boll weevil.
Nov., 1917	George Ellery Hale oversees the installation of the Hooker Telescope on Mount Wilson.
Jan. 8, 1918	Harlow Shapley proves the Sun is distant from the center of our galaxy.

1919   Francis William Aston builds the first mass spectrograph and discovers isotopes.

1919   Richard von Mises develops the frequency theory of probability.

1919   The principles of shortwave radio communication are discovered.

1919-1921   Vilhelm Bjerknes discovers fronts in atmospheric circulation.

Spring, 1919   Karl von Frisch discovers that bees communicate through body movements.

Nov. 6, 1919   Albert Einstein's theory of gravitation is confirmed.

Early 1920's   Vesto Slipher presents evidence of redshifts in galactic spectra.

1920-1930   Robert Andrews Millikan names cosmic rays and investigates their absorption.

Dec. 13, 1920   Albert A. Michelson measures the diameter of a star.

1921   John A. Larson constructs the first modern polygraph.

1921   Albert Calmette and Camille Guérin develop the tuberculosis vaccine BCG.

1921   Emmy Noether publishes the theory of ideals in rings.

1921-1923   William Grant Banting and J. J. R. Macleod win the Nobel Prize for the discovery of insulin.

1922   Elmer McCollum names vitamin D and pioneers its use against rickets.

Nov. 4, 1922   Howard Carter discovers the tomb of Tutankhamen.

1923   Arthur Holly Compton discovers the wavelength change of scattered X rays.

1923   Roy Chapman Andrews discovers the first fossilized dinosaur eggs.

1923   Vladimir Zworykin develops an early type of television.

1923   Louis de Broglie introduces the theory of wave-particle duality.

1923-1951   Reuben Leon Kahn develops a modified syphilis test and the universal serologic test.

Summer, 1923   Otto Zdansky discovers Peking man.

1924   Harry Steenbock discovers that sunlight increases vitamin D in food.

1924   Theodor Svedberg develops the ultracentrifuge.

Mar., 1924   Arthur Stanley Eddington formulates the mass-luminosity law for stars.

Summer, 1924   Raymond Arthur Dart discovers the first recognized australopithecine fossil.

Dec., 1924   Edwin Powell Hubble determines the distance to the Andromeda nebula and demonstrates that other galaxies are independent systems.

1925   Fred Whipple finds iron to be an important constituent of red blood cells.

Spring, 1925   Wolfgang Pauli formulates the exclusion principle.

Apr., 1925-May, 1927   The German *Meteor* expedition discovers the Mid-Atlantic Ridge.

Mar. 16, 1926   Robert Goddard launches the first liquid fuel propelled rocket.

July, 1926   Arthur Stanley Eddington publishes *The Internal Constitution of the Stars*.

Aug., 1926-Sept., 1928   Warner Bros. introduces talking motion pictures.

1927   Georges Lemaître proposes the big bang theory.

1927   Jan Hendrik Oort proves the spiral structure of the Milky Way.

Feb.-Mar., 1927   Werner Heisenberg articulates the uncertainty principle.

May 20-21, 1927   Charles Lindbergh makes the first nonstop solo flight across the Atlantic Ocean.

1928   Vannevar Bush builds the first differential analyzer.

1928   George Gamow explains radioactive alpha-decay with quantum tunneling.

1928-1932   Albert Szent-Györgyi discovers vitamin C.

Jan., 1928   George N. Papanicolaou develops the Pap test for diagnosing uterine cancer.

Aug., 1928   Margaret Mead publishes *Coming of Age in Samoa*.

Sept., 1928   Alexander Fleming discovers penicillin in molds.

1929   Edwin Powell Hubble confirms the expanding universe.

Apr. 22, 1929	Hans Berger develops the electroencephalogram (EEG).
July, 1929	Philip Drinker and Louis Shaw develop an iron lung mechanical respirator.
July, 1929-July, 1931	Kurt Gödel proves incompleteness-inconsistency for formal systems, including arithmetic.
Winter, 1929-1930	Bernhard Voldemar Schmidt invents the corrector for the Schmidt camera and telescope.
1930	Construction begins on the Empire State Building.
1930	Thomas Midgley introduces dichlorodifluoromethane as a refrigerant gas.
1930	Hans Zinsser develops an immunization against typhus.
1930	Bernard Lyot builds the coronagraph for telescopically observing the Sun's outer atmosphere.
1930-1931	Linus Pauling develops his theory of the chemical bond.
1930-1932	Karl G. Jansky's experiments lead to the founding of radio astronomy.
1930-1935	Edwin H. Armstrong perfects FM radio.
Feb. 18, 1930	Clyde Tombaugh discovers Pluto.
Jan. 2, 1931	Ernest Orlando Lawrence develops the cyclotron.
Apr., 1931	Ernst Ruska creates the first electron microscope.
May 27, 1931	Auguste Piccard travels to the stratosphere by balloon.
1931-1935	Subramanyan Chandrasekhar calculates the upper limit of a white dwarf star's mass.
Feb., 1932	James Chadwick discovers the neutron.
Apr., 1932	John Douglas Cockcroft and Ernest Walton split the atom with a particle accelerator.
Sept., 1932	Carl David Anderson discovers the positron.
1932-1935	Gerhard Domagk discovers that a sulfonamide can save lives.
Nov., 1933	Enrico Fermi proposes the neutrino theory of beta decay.
1933-1934	Frédéric Joliot and Irène Joliot-Curie develop the first artificial radioactive element.
1934	Ruth Benedict publishes *Patterns of Culture*.

1934	Pavel Cherenkov discovers the Cherenkov effect.
1934	Fritz Zwicky and Walter Baade propose their theory of neutron stars.
1934-1938	Chester F. Carlson invents xerography.
Aug. 11-15, 1934	William Beebe and Otis Barton set a diving record in a bathysphere.
Fall, 1934	John H. Gibbon develops the heart-lung machine.
Nov., 1934-Feb., 1935	Hideki Yukawa proposes the existence of mesons.
1935	Robert Alexander Watson-Watt and associates develop the first radar.
1935	Sydney Chapman determines the lunar atmospheric tide at moderate latitudes.
1935-1936	Alan M. Turing invents the Universal Turing Machine.
Jan., 1935	Charles F. Richter develops a scale for measuring earthquake strength.
Feb., 1935-Oct., 1938	Wallace Carothers patents nylon.
Nov.-Dec., 1935	Antonio Egas Moniz develops prefrontal lobotomy.
1936	Inge Lehmann discovers the Earth's inner core.
1936	Erwin Wilhelm Müller invents the field emission microscope.
Mar. 1, 1936	The completion of Boulder Dam creates Lake Mead, the world's largest reservoir.
Nov. 23, 1936	Fluorescent lighting is introduced.
1937	Max Theiler introduces a vaccine against yellow fever.
1937	Ugo Cerletti and Lucino Bini develop electroconvulsive therapy for treating schizophrenia.
Jan.-Sept., 1937	Emilio Segrè identifies the first artificial element, technetium.
Mar., 1937	Hans Adolf Krebs describes the citric acid cycle.
June-Sept., 1937	Grote Reber builds the first radio telescope.
Fall, 1937-Winter, 1938	Franz Weidenreich reconstructs the face of Peking man.
1938	George S. Callendar connects industry with increased atmospheric carbon dioxide.
1938	Albert Hofmann synthesizes the potent psychedelic drug LSD-25.

1938	Peter Kapitsa explains superfluidity.
Dec., 1938	Otto Hahn splits an atom of uranium.
1939	The Bourbaki group publishes *Éléments de mathématique*.
1939	Paul Hermann Müller discovers that DDT is a potent insecticide.
Feb. 15, 1939	J. Robert Oppenheimer calculates the nature of black holes.
Early 1940's	A secret English team develops Colossus.
Late 1940's	Willard F. Libby introduces the carbon-14 method of dating ancient objects.
1940	The first color television broadcast takes place.
May, 1940	Baron Florey and Ernst Boris Chain develop penicillin as an antibiotic.
Sept. 12, 1940	Seventeen-thousand-year-old paintings are discovered in Lascaux cave.
Feb. 23, 1941	Glenn Seaborg and Edwin McMillan make element 94, plutonium.
May 15, 1941	The first jet plane using Frank Whittle's engine is flown.
1942-1947	Grote Reber makes the first radio maps of the universe.
Dec. 2, 1942	Enrico Fermi creates the first controlled nuclear fission chain reaction.
1943-1944	Oswald Avery, Colin Macleod, and Maclyn McCarty determine that DNA carries hereditary information.
1943-1944	Carl Friedrich von Weizsäcker finalizes his quantitative theory of planetary formation.
1943-1946	John Presper Eckert and John William Mauchly develop the ENIAC computer.
Spring, 1943	Jacques Cousteau and Émile Gagnan develop the Aqua-Lung.
Sept., 1943-Mar., 1944	Selman Abraham Waksman discovers the antibiotic streptomycin.
Nov. 4, 1943	The world's first nuclear reactor is activated.
1944	The Germans use the V-1 flying bomb and the V-2 goes into production.
Jan., 1944	Gerard Peter Kuiper discovers that Titan has an atmosphere.

Nov., 1944	Alfred Blalock and Helen Taussig perform the first "blue baby" operation.
1944-1949	Dorothy Crowfoot Hodgkin solves the structure of penicillin.
1944-1952	Sir Martin Ryle's radio telescope locates the first known radio galaxy.
1945	Benjamin Minge Duggar discovers aureomycin, the first of the tetracyclines.
Jan., 1945	Artificial fluoridation of municipal water supplies to prevent dental decay is introduced.
July 16, 1945	The first atomic bomb is successfully detonated.
July 12, 1946	Vincent Joseph Schaefer performs cloud seeding by using dry ice.
Nov., 1946	University of California physicists develop the first synchrocyclotron.
1947	Dennis Gabor develops the basic concept of holography.
1947	Willis Eugene Lamb, Jr., and Robert C. Retherford discover the lambshift.
Spring, 1947	Archaeologists unearth ancient Dead Sea scrolls.
Nov.-Dec., 1947	William Shockley, John Bardeen, and Walter Brattain discover the transistor.
1948	George Gamow and associates develop the big bang theory.
1948	The steady-state theory of the universe is advanced by Hermann Bondi, Thomas Gold, and Fred Hoyle.
1948	Eric Jacobsen introduces a drug for the treatment of alcoholism.
June 3, 1948	George Ellery Hale constructs the largest telescope of the time.
Nov. 26, 1948	Edwin Herbert Land invents a camera/film system that develops instant pictures.
1949	X rays from a synchrotron are first used in medical diagnosis and treatment.
Feb. 24, 1949	The first rocket with more than one stage is created.
Aug., 1949	BINAC, the first electronic stored-program computer, is completed.
1950's	Robert Wallace Wilkins discovers Reserpine, the first tranquilizer.

1950's	Choh Hao Li isolates the human growth hormone.
1950	The artificial sweetener cyclamate is introduced.
1950	William Clouser Boyd defines human races by blood groups.
Mid-1950's	Severo Ochoa creates synthetic RNA.
1951	Robert Hofstadter discovers that protons and neutrons each have a structure.
1951	Fritz Albert Lipmann discovers acetyl coenzyme a.
1951	UNIVAC I becomes the first commercial electronic computer and the first to use magnetic tape.
1951-1952	Edward Teller and Stanislaw Ulam develop the first hydrogen bomb.
1951-1953	James Watson and Francis Crick develop the double-helix model for DNA.
May, 1951-May, 1954	Jan Hendrik Oort postulates the existence of the Oort Cloud.
Dec. 20, 1951	The world's first breeder reactor produces electricity while generating new fuel.
1952	Eugene Aserinsky discovers rapid eye movement (REM) in sleep and dreams.
Feb. 23, 1952	Douglas Bevis describes amniocentesis as a method for disclosing fetal genetic traits.
July 2, 1952	Jonas Salk develops a polio vaccine.
Aug., 1952	Walter Baade corrects an error in the cepheid luminosity scale.
1952-1956	Erwin Wilhelm Müller develops the field ion microscope.
1953	Vincent du Vigneaud synthesizes oxytocin, the first peptide hormone.
1953	Stanley Miller reports the synthesis of amino acids.
1953	Gérard de Vaucouleurs identifies the local supercluster of galaxies.
1953-1959	The liquid bubble chamber is developed.
1954-1957	John Backus's IBM team develops the FORTRAN computer language.
Apr. 30, 1954	Elso Barghoorn and Stanley Tyler discover 2-billion-year-old microfossils.

May, 1954 Bell Telephone scientists develop the photovoltaic cell.

1955 Kenneth Franklin and Bernard Burke discover radio emissions from Jupiter.

1955 Sir Martin Ryle constructs the first radio interferometer.

1956 The first transatlantic telephone cable is put into operation.

1956 Bruce Heezen and Maurice Ewing discover the midoceanic ridge.

Apr.-Dec., 1956 Birth control pills are tested in Puerto Rico.

1957 Albert Bruce Sabin develops an oral polio vaccine.

1957 Alick Isaacs and Jean Lindenmann discover interferons.

1957 Sony develops the pocket-sized transistor radio.

Feb. 7, 1957 John Bardeen, Leon N. Cooper, and John Robert Schrieffer explain superconductivity.

Aug., 1957 The Jodrell Bank radio telescope is completed.

Oct. 4, 1957 The Soviet Union launches the first artificial satellite, Sputnik 1.

Oct. 11, 1957 Leo Esaki demonstrates electron tunneling in semiconductors.

Dec. 2, 1957 The United States opens the first commercial nuclear power plant.

1958 Frederick Sanger wins the Nobel Prize for the discovery of the structure of insulin.

1958 James Van Allen discovers the Earth's radiation belts.

1958 Ian Donald is the first to use ultrasound to examine unborn children.

Jan. 2, 1958 Eugene N. Parker predicts the existence of the solar wind.

Jan. 31, 1958 The United States launches its first orbiting satellite, Explorer 1.

1959 Grace Hopper invents the computer language COBOL.

1959 A corroded mechanism is recognized as an ancient astronomical computer.

1959	A radio astronomy team sends and receives radar signals to and from the Sun.
June 26, 1959	The St. Lawrence Seaway is opened.
July 17, 1959	Louis and Mary Leakey find a 1.75-million-year-old fossil hominid.
Sept. 13, 1959	Luna 2 becomes the first human-made object to impact on the Moon.
Early 1960's	The plastic IUD is introduced for birth control.
Early 1960's	Anthropologists claim that Ecuadorian pottery shows transpacific contact in 3000 B.C.E.
Early 1960's	Roger Sperry discovers that each side of the brain can function independently.
1960	The Mössbauer effect is used in the detection of gravitational redshifting.
1960	Scientists develop a technique to date ancient obsidian.
1960-1962	Harry Hammond Hess concludes the debate on continental drift.
1960-1969	A vaccine is developed for German measles.
Spring, 1960	Juan Oró detects the formation of adenine from cyanide solution.
Apr. 1-June 14, 1960	Tiros 1 becomes the first experimental weather reconnaissance satellite.
July, 1960	The first laser is developed in the United States.
Aug. 12, 1960	Echo, the first passive communications satellite, is launched.
1961	Frank L. Horsfall announces that cancer results from alterations in the DNA of cells.
1961	Marshall Nirenberg invents an experimental technique that cracks the genetic code.
Apr. 12, 1961	Yuri Gagarin becomes the first human to orbit Earth.
May 5, 1961	Alan Shepard is the first United States astronaut in space.
Dec., 1961	Melvin Calvin identifies the chemical pathway of photosynthesis.
1962	Lasers are used in eye surgery for the first time.
1962	John Glenn is the first American to orbit Earth.

1962	Riccardo Giacconi and associates discover the first known X-ray source outside the solar system.
July 10, 1962	Telstar, the first commercial communications satellite, relays live transatlantic television pictures.
Aug., 1962-Jan., 1963	Mariner 2 becomes the first spacecraft to study Venus.
Sept. 27, 1962	Rachel Carson publishes *Silent Spring*.
1962-1967	Colin Renfrew, J. E. Dixon, and J. R. Cann reconstruct ancient Near Eastern trade routes.
1963	The cassette for recording and playing back sound is introduced.
1963	Maarten Schmidt makes what constitutes the first recognition of a quasar.
1963	Paul J. Cohen shows that Georg Cantor's continuum hypothesis is independent of the axioms of set theory.
1963-1965	Arno Penzias and Robert Wilson discover cosmic microwave background radiation.
1964	Quarks are postulated by Murray Gell-Mann and George Zweig.
1964-1965	John G. Kemeny and Thomas E. Kurtz develop the BASIC computer language.
1964-1965	Richard Rayman Doell and Brent Dalrymple discover the magnetic reversals of Earth's poles.
Nov. 21, 1964	The Verrazano Bridge opens.
1965	The Sealab 2 expedition concludes.
Mar. 18, 1965	The first spacewalk is conducted from Voskhod 2.
Nov. 16, 1965-Mar. 1, 1966	Venera 3 is the first spacecraft to impact on another planet.
Dec., 1965	The orbital rendezvous of Gemini 6 and 7 succeeds.
1966	Robert Ardrey's *The Territorial Imperative* argues that humans are naturally territorial.
Jan., 1966	Elwyn L. Simons identifies a 30-million-year-old primate skull.
Jan. 31-Feb. 8, 1966	The Soviet Luna 9 makes the first successful lunar soft landing.
Aug. 10-Oct. 29, 1966	The Lunar Orbiter 1 sends photographs of the Moon's surface.

1967   Rene Favaloro develops the coronary artery bypass operation.

1967   Syurkuro Manabe and Richard Wetherald warn of the greenhouse effect and global warming.

1967   Raymond Davis constructs a solar neutrino detector.

1967   Benoît Mandelbrot develops non-Euclidean fractal measures.

1967-1968   Elso Barghoorn and coworkers find amino acids in 3-billion-year-old rocks.

Aug.-Sept., 1967   Arthur Kornberg and coworkers synthesize biologically active DNA.

Nov., 1967-Feb., 1968   Jocelyn Bell discovers pulsars, the key to neutron stars.

Dec., 1967   Christiaan Barnard performs the first human heart transplant.

1968   Jerome I. Friedman, Henry W. Kendell, and Richard E. Taylor discover quarks.

1968   The *Glomar Challenger* obtains thousands of ocean floor samples.

1968   John Archibald Wheeler names the phenomenon "black holes."

1969   Bubble memory devices are created for use in computers.

1969   The Soyuz 4 and 5 spacecraft dock in orbit.

1969-1970   The first jumbo jet service is introduced.

1969-1974   Very long baseline interferometry is developed for high-resolution astronomy and geodesy.

July 20, 1969   Neil Armstrong and Edwin "Buzz" Aldrin land on the Moon.

Dec. 10, 1969   Derek H. R. Barton and Odd Hassel share the Nobel Prize for determining the three-dimensional shapes of organic compounds.

1970   The floppy disk is introduced for storing data used by computers.

Nov. 10, 1970-Oct. 1, 1971   Lunokhod 1 lands on the Moon.

1971   Direct transoceanic dialing begins.

1971   The microprocessor "computer on a chip" is introduced.

1971-1972	Mariner 9 is the first known spacecraft to orbit another planet.
May 19, 1971-Mar., 1972	Mars 2 is the first spacecraft to impact on Mars.
Mar., 1972	Pioneer 10 is launched.
Apr., 1972	Godfrey Hounsfield introduces a cat scanner that can see clearly into the body.
Sept., 1972	Murray Gell-Mann formulates the theory of quantum chromodynamics (qcd).
Sept., 1972	Texas Instruments introduces the first commercial pocket calculator.
Sept. 23, 1972	David Janowsky publishes a cholinergic-adrenergic hypothesis of mania and depression.
Dec. 31, 1972	The United States government bans DDT use to protect the environment.
1973	Stanley Cohen and Herbert Boyer develop recombinant DNA technology.
Feb., 1973-Mar., 1974	Organic molecules are discovered in Comet Kohoutek.
May 14, 1973-Feb. 8, 1974	Skylab inaugurates a new era of space research.
Nov. 3, 1973-Mar. 24, 1975	Mariner 10 is the first mission to use gravitational pull of one planet to help it reach another.
Dec., 1973-June, 1974	F. Sherwood Rowland and Mario J. Molina theorize that ozone depletion is caused by Freon.
Feb., 1974	Howard Georgi and Sheldon Glashow develop the first grand unified theory.
Apr., 1974	Optical pulses shorter than one trillionth of a second are produced.
June, 1974	Tunable, continuous wave visible lasers are developed.
Aug.-Sept., 1974	The J/psi subatomic particle is discovered.
Nov., 1974	Donald Johansen and Tim White discover "Lucy," an early hominid skeleton.
Oct. 22, 1975	Soviet Venera spacecraft transmit the first pictures from the surface of Venus.
1976	Thomas Kibble proposes the theory of cosmic strings.
July 20-Sept. 3, 1976	Viking spacecraft send photographs to Earth from the surface of Mars.

1977	Alan J. Heeger and Alan G. MacDiarmid discover that iodine-doped polyacetylene conducts electricity.
1977	Deep-sea hydrothermal vents and new life-forms are discovered.
Mar. 10-11, 1977	Astronomers discover the rings of the planet Uranus.
Apr., 1977	Apple II becomes the first successful preassembled personal computer.
May, 1977	The first commercial test of fiber-optic telecommunications is conducted.
Sept. 16, 1977	Andreas Gruentzig uses percutaneous transluminal angioplasty, via a balloon catheter, to unclog diseased arteries.
Sept., 1977-Sept., 1989	Voyager 1 and 2 explore the planets.
July 25, 1978	Louise Brown gives birth to the first "test-tube" baby.
1978-1981	Heinrich Rohrer and Gerd Binnig invent the scanning tunneling microscope.
Mar. 4-7, 1979	The first ring around Jupiter is discovered.
Aug., 1979	An ancient sanctuary is discovered in El Juyo Cave, Spain.
1980	Paul Berg, Walter Gilbert, and Frederick Sanger develop techniques for genetic engineering.
1980	Evidence is found of a worldwide catastrophe at the end of the Cretaceous period.
1980	The inflationary theory solves long-standing problems with the big bang.
Jan. 14, 1980	Robert Louis Griess constructs "The Monster," the last sporadic group.
Feb. 5, 1980	Klaus von Klitzing discovers the quantized Hall effect.
May, 1980	Pluto is found to possess a thin atmosphere.
June, 1980	Radar observations show that Mayan agricultural centers are surrounded by canals.
1981	The U.S. Centers for Disease Control recognizes AIDS for the first time.
1981-1982	A human growth hormone gene transferred to a mouse creates giant mice.

May-June, 1981   Bell Laboratories scientists announce a liquid-junction solar cell of 11.5 percent efficiency.

June, 1981   Joseph Patrick Cassinelli and associates discover R136a, the most massive star known at the time.

Aug. 12, 1981   The IBM personal computer, using DOS, is introduced.

Sept., 1981   William H. Clewell corrects hydrocephalus by surgery on a fetus.

Nov. 12-14, 1981   *Columbia*'s second flight proves the practicality of the space shuttle.

1982   Thomas Cech and Sidney Altman demonstrate that RNA can act as an enzyme.

1982   William Castle Devries implants the first Jarvik-7 artificial heart.

1982   Étienne-Émile Baulieu develops RU-486, a pill that induces abortion.

1982-1983   Compact disc players are introduced.

1982-1983   Fernand Daffos uses blood taken through the umbilical cord to diagnose fetal disease.

1982-1989   Astronomers discover an unusual ring system of the planet Neptune.

Apr., 1982   Solar One, the prototype power tower, begins operation.

May 14, 1982   The first commercial genetic engineering product, Humulin, is marketed by Eli Lilly.

1983   The artificial sweetener aspartame is approved for use in carbonated beverages.

1983   Carlo Rubbia and Simon van der Meer isolate the intermediate vector bosons.

Jan.-Oct., 1983   The first successful human embryo transfer is performed.

Mar. 8, 1983   IBM introduces a personal computer with a standard hard disk drive.

Apr. 4, 1983   The first tracking and data-relay satellite system opens a new era in space communications.

Sept., 1983   Andrew Murray and Jack Szostak create the first artificial chromosome.

Nov. 28, 1983   *Spacelab* 1 is launched aboard the space shuttle.

1984   Optical disks for the storage of computer data are introduced.

1984   Charles Gald Sibley and Jon Ahlquist discover a close human and chimpanzee genetic relationship.

1984   Steen M. Willadsen clones sheep using a simple technique.

1985   The British Antarctic Survey confirms the first known hole in the ozone layer.

1985   Construction of the world's largest land-based telescope, the Keck, begins in Hawaii.

Mar. 6, 1985   Alec Jeffreys discovers the technique of genetic fingerprinting.

Oct., 1985   The Tevatron particle accelerator begins operation at Fermilab.

1986-1987   R. Brent Tully discovers the Pisces-Cetus supercluster complex.

Jan., 1986   J. Georg Bednorz and Karl Alexander Müller discover a high-temperature superconductor.

Feb. 20, 1986   The first permanently manned space station is launched.

Apr. 26, 1986   The Chernobyl nuclear reactor explodes.

July, 1986   A genetically engineered vaccine for hepatitis B is approved for use.

Oct., 1986   A gene that can suppress the cancer retinoblastoma is discovered.

Dec. 14-23, 1986   Burt Rutan and Chuck Yeager pilot the *Voyager* around the world without refueling.

Feb. 23, 1987   Supernova 1987a corroborates the theories of star formation.

Sept., 1987   Wade Miller discovers a dinosaur egg containing the oldest known embryo.

1987-1988   Scientists date a *Homo sapiens* fossil at ninety-two thousand years.

1988   Henry Erlich develops DNA fingerprinting from a single hair.

Apr. 24, 1990   NASA launches the Hubble Space Telescope.

1990's-2002 Particle physicists demonstrate that neutrinos—atomic particles long thought to be without mass—do indeed have mass and that they can change "flavor."

1992 China revives the Cold War threat of international nuclear war with the explosion of one of the most powerful nuclear devices ever tested.

1992 More than ten thousand scientists and AIDS (acquired immune deficiency syndrome) activists meeting in Amsterdam reveal the possibility of a new AIDS-like virus.

1993 Andrew Wiles presents his proof of the "Last Theorem" of Pierre de Fermat, which had defied solution by mathematicians for more than three and a half centuries.

1993 In the most dramatic report of ozone depletion since the phenomenon was first reported, the World Meteorological Organization announces a rapid decline in ozone levels in the Northern Hemisphere.

1994 Astronomers use the Hubble Space Telescope to find evidence for the existence of a black hole in the center of galaxy M87.

1994 The Hubble Space Telescope provides astronomers with clear images of distant objects in the universe.

1994 Colombia, a major exporter of cocaine to the United States, legalizes the possession and private use of small amounts of cocaine, marijuana, and some other drugs for its citizens, provoking anger among U.S. drug enforcement officials.

1994 An international conference sponsored by the United Nations emphasizes links among population control, economic development, and the advancement of women.

1995 National and international health organizations react quickly to contain an outbreak of the deadly Ebola virus in Kikwit, Zaire.

1995 After sixty years' absence, wolves are restored to Yellowstone National Park in the western United States under a provision of the Endangered Species Act.

1995   The Kobe, Japan, earthquake of January 17, 1995, kills 5,500 people, injures 37,000, and does damage exceeding $50 billion, one of the most costly natural disasters on record.

1995   Two teams of physicists announce the discovery of the top quark, the last of six such subatomic particles predicted by scientific theory.

1995   U.S. astronauts aboard the shuttle *Atlantis* dock with the space station Mir on a mission that sets the stage for future rendezvous and construction of an international space station.

1995   At 12:01 A.M. on August 24, 1995, the first copy of Microsoft Windows 95 is sold. Windows 95, which makes using an Intel personal computer easy and intuitive, becomes the operating system of choice for personal computers and one of the most successful software products ever developed.

1995   The second assessment report of the Intergovernmental Panel on Climate Change (IPCC) projects a rise in global mean surface temperatures. The rise would constitute the fastest rate of change since the end of the last Ice Age.

1995   Six years after its deployment from the space shuttle *Atlantis* I, the Galileo spacecraft reaches its destination, the planet Jupiter.

1996   Dolly the sheep, is born. She is the first vertebrate cloned from the cell of an adult vertebrate.

1996   Scientists at the National Aeronautics and Space Administration (NASA) find traces of life processes and possible microscopic fossils in a meteorite believed to have come from Mars.

1997   Physicists at the Massachusetts Institute of Technology announce the success of an elementary version of a laser that produces a beam of atoms rather than a beam of light.

1997   After a six-year journey through interplanetary space, the Galileo spacecraft passes within 370 miles of Jupiter's moon Europa, revealing an ice-enshrouded world whose surface characteristics suggest an underlying planetary ocean that may harbor extraterrestrial life.

1997   Anthropologists discover the fossil skull of a boy who lived in Spain nearly 800,000 years ago. His skull combines features of both modern humans and earlier human species.

1997   A spacecraft that was launched from the Kennedy Space Center in Florida on December 4, 1996, lands safely on Mars after a flight lasting seven months.

1997   The discovery of ancient Roman shipwrecks in the Mediterranean Sea confirms the theory that ancient sailors did not simply hug the coast as they were engaging in trade across the Mediterranean.

1997   Scientists discover ancient fossil footprints left by a woman who walked on the shores of Langebaan Lagoon, South Africa, 117,000 years ago.

1998   Scientists announce that preliminary findings from the Lunar Prospector mission suggest the presence of water ice in the shadowed craters near the Moon's poles.

1998   The Monahans meteorite is the first extraterrestrial object to provide a sample of liquid water from an asteroid. The water, trapped in salt crystals, demonstrates that liquid water existed early in the history of the solar system, and the association of water with salt crystals suggests that brine evaporated on or near the surface of the asteroid.

1998   Developed by a team of scientists of the Pfizer Company, Viagra is the first anti-impotence drug to be approved by the U.S. Food and Drug Administration.

1998   Despite condemnation from the proponents of nuclear nonproliferation, India and Pakistan test nuclear weapons and join those countries possessing nuclear weapons.

1998   The annual ozone hole extends a record 10.5 million square miles (27.3 million square kilometers).

1998   The first digital high-definition television signals are broadcast.

1999   A team of surgeons perform a successful hand transplant operation in Louisville, Kentucky, enabling the recipient to perform twisting and gripping functions and to feel sensation in the hand.

1999   Scientists trace HIV, the virus that causes AIDS, to chimpanzees.

1999   In an effort to produce alternative sources of clean energy, researchers generate nuclear energy on a tabletop by both fusion and fission.

1999   A Danish physicist and her collaborators reduce light's speed from 186,000 miles (299,274 kilometers) per second to 38 miles (61 kilometers) per hour.

1999   Careful analysis of light from Upsilon Andromedae reveals the first known multiple-planet system orbiting a normal star.

1999   A team of scientists at the Lawrence Berkeley Laboratory Nuclear Science Division detect the formation of two new elements, with atomic numbers 116 and 118, as the result of bombarding lead targets with krypton ions in the 88-inch (2.2-meter) cyclotron.

1999   Continuous occupation of the Russian Mir Space Station ends in August, 1999, and Mir falls out of Earth orbit in 2001.

1999   Physicists produce nickel-48, the most proton-rich nucleus, an international breakthrough in nuclear physics.

1999   The Mars Climate Orbiter, a $125 million robotic spacecraft designed to investigate weather on Mars, disappears as it is about to enter orbit around Mars. The error is later traced to human miscalculation.

1999   Researchers announce the identification of an enzyme that plays a key role in the development of Alzheimer's disease.

1999   A team of scientists at Brown University present topographical measurements that indicate that an ocean once existed on Mars.

Oct. 12, 1999   According to United Nations data, the world's six billionth person is born.

Dec., 1999   Seven astronauts aboard the space shuttle *Discovery* successfully restore the Hubble Space Telescope to operation.

2000   Intel introduces its Pentium 4 microprocessor, with processing speeds of 1.5 gigahertz, a vast improvement over its first microprocessor, the Intel 8088, introduced in 1979 for the IBM personal computer.

2000   Imaging radar aboard the space shuttle *Endeavour* captures data to assemble the most comprehensive topographic map of Earth, covering 80 percent of its land surface.

2000   The Framingham study, which has followed thousands of women throughout their lives, reveals that hormone-replacement therapy (HRT) might not prevent coronary disease in postmenopausal women, as previously thought; most physicians immediately order their patients on HRT to diminish or cease their dosages.

2000   Scholars from two research institutions announce the discovery of at least nine planets around stars other than the Sun, bringing the total number of known extrasolar planets to at least fifty; by 2005, the number of known extrasolar planets has exceeded one hundred.

2000   The Food and Drug Administration approves medical abortions using mifepristone (RU-486) as an alternative to surgical abortion.

2000   The first construction begins on the International Space Station, a structure for scientific and biological research to be erected in Earth orbit.

2000   The gas-electric "hybrid" automobile is brought to market.

2000   With the aid of computers, geneticists are rapidly sequencing the genomes of many organisms, culminating in the year's sequencing of the complete genome *Drosophila melanogaster*, the fruit fly.

Feb. 14, 2000   The Near Earth Asteroid Rendezvous (NEAR) spacecraft begins a yearlong orbit of the asteroid Eros, gathering data on its chemical composition, mineralogy, shape, and structure.

2001   Scientists advance a new area of applied science, nanoelectronics, by assembling molecules into basic circuits.

Feb. 10, 2001   The human genome is completely sequenced, opening a new era of medical promise; the event marks the most important breakthrough in genetics since the discovery of the double-helical structure of DNA in 1953.

2002   A variety of small RNA molecules are discovered to be capable of altering gene expression and even the genome itself.

2002-2003   Researchers announce genetic predispositions toward depression and bipolar disorder.

2003   Biophysicists experiment with "quantum dots," tiny semiconductor nanocrystals that glow in the presence of laser light, to enhance biological imaging techniques.

2003   Biologists discover that mouse stem cells can develop into both sperm and egg cells in vitro, raising the question of whether the same is possible with human stem cells.

2003   Physicists confirm the existence of "left-handed" materials, which have a negative refractive index (they bend light at a negative angle when it passes into them from a different medium) as well as other odd and potentially useful properties.

2003   The combination of conventional chemotherapy and new antiangiogenesis drugs—which starve cancer tumors of their blood supply by preventing them from growing blood vessels—proves effective with colon cancer patients.

2003   The Wilkinson Anisotropy Microwave Probe maps the universe showing the cosmic background radiation, the "afterglow" of the big bang, and pinpoints the age of the universe at 13.7 billion years.

Jan., 2004   The Mars Exploration Rovers, Spirit and Opportunity, land at different locations on the Martian surface and return unprecedented photographs of topographic features as well as geological data.

Dec. 26, 2004   A devastating tsunami, generated by an earthquake in the ocean near Indonesia, kills more than 200,000 in nations from Sri Lanka to Indonesia.

Jan. 27, 2005   Oxford University's climateprediction.net project announces evidence of a long-term increase in Earth's surface temperature in the range of 2° to 11° Celsius as a result of global warming.

Feb. 17, 2005   Two human skulls discovered in Ethiopia by Richard Leakey in 1967 are redated to 195,000 years old, the oldest known remains of modern human beings.

July 4, 2005   The Deep Impact spacecraft reaches Comet Tempel 1 and launches a 372-kilogram copper projectile into the comet's icy surface to collect data.

July-September, 2005   Xena, a body beyond Pluto that orbits the Sun, is discovered by astronomers at the University of Hawaii's Keck Observatory in July and its moon Gabrielle is discovered in September. The question of Xena's planetary status—like those of several other trans-Neptunian objects discovered since 1995—is debated by astronomers.

September, 2005   The U.S. National Snow and Ice Data Center and the National Aeronautics and Space Administration report "a stunning reduction" in Arctic sea ice, 20 percent below the mean average during Septembers from 1978 to 2001.

WEB SITES

This appendix includes a wide variety of Web sites that range from providing basic information to exploring specific subjects in depth. Several of the sites are interactive; many provide additional links to more resources in their respective subject areas. Although URLs change over time, they are often rerouted to new Web addresses automatically; alternatively, a search on the sponsor listed will usually take the user to the site in question if the Web address has changed.

GENERAL SITES: EDUCATION, NEWS, REFERENCE

Bill Nye the Science Guy's Nye Labs Online
http://nyelabs.kcts.org

Conversion Factors
http://www.wsdot.wa.gov/Metrics/factors.htm

Dictionary of Scientific Quotations
http://naturalscience.com/dsqhome.html

Discovery Channel
http://discover.com

Eisenhower National Clearinghouse Resources Finder
http://www.enc.org/rf/nf_index.htm#rf

Electronic Journal of Science Education
http://unr.edu/homepage/jcannon/ejse/ejse.html

Environment News Network (ENN)
http://www.enn.com

How Stuff Works
http://www.howstuffworks.com

Internet Public Library Science and Technology Resources
http://www.ipl.org/ref/rr/static/scioo.oo.oo.html

Journal of Young Investigators
http://www.jyi.org

Library of Congress Science Reading Room
http://lcweb.loc.gov/rr.scitech

Livescience.com
http://www.livescience.com

MagPortal.com: Magazine Articles on Science and Technology
http://www.MagPortal.com/c/sci

National Academy of Sciences
http://www.nasonline.org/site/PageServer

National Aeronautics and Space Administration (NASA)
http://education.nasa.gov

National Geographic Society
http://www.nationalgeographic.com

National Institute for Science Education
http://www.wcer.wisc.edu/nise

National Science Foundation
http://www.nsf.gov

National Science Teachers Association
http://www.nsta.org

Nature Magazine
http://www.nature.com

New Scientist
http://www.newscientist.com

Newton's Apple Index
http://ericir.syr.edu/Projects/Newton

Nova (Public Broadcasting Service)
http://www.pbs.org/wgbh/nova

Odyssey Adventures in Science
http://www.odysseymagazine.com

On Being a Scientist: Responsible Conduct in Research
http://www.nap.edu/readingroom/books/obas

Popular Science Magazine
http://www.popularscience.com

Resources in Science and Engineering Education
http://www2.ncsu.edu/unity/lockers/users/f/felder/public/RMF.html

Scholarly Societies Project
http://www.lib.uwaterloo.ca/society/overview.html

Science Learning Network (SLN)
http://www.sln.org

Science Magazine
http://www.sciencemag.org/content/vol309/issue5743/twis.shtml

Science/Nature for Kids
http://kidscience.about.com

Science News
http://www.sciencenews.org

Science Online
http://www.scienceonline.org

Science Service Historical Image Collection
http://americanhistory.si.edu/scienceservice

Scientific American
http://www.sciam.com

Scout Report for Science and Engineering
http://scout.cs.wisc.edu/

Society for College Science Teachers
http://science.clayton.edu/scst

Statistical Reports of U.S. Science and Engineering
http://www.nsf.gov/sbe/srs/stats.htm

Temperature Conversion Calculator
http://www.cchem.berkeley.edu/ChemResources/temperature.html

U.S. House of Representatives Committee on Science
http://www.house.gov/science/welcome.htm

U.S. Senate Committee on Commerce, Science and Transportation
http://www.senate.gov/~commerce

Why Files: The Science Behind the News
http://whyfiles.news.wisc.edu

Yahoo! Science and Technology News
http://dailynews.yahoo.com

ARCHAEOLOGY

Ancient Technologies and Archaeological Materials
http://www.uiuc.edu/unit/ATAM/index.html

Archaeology Magazine Online
http://www.archaeology.org/main.html

Web Info Radiocarbon Dating
http://www.c14dating.com

WWWorld of Archaeology
http://www.archaeology.org/wwwarky/wwwarky.html

ASTRONOMY AND AEROSPACE

About.com Aerospace and Astronomy Sites
http://space.about.com

American Meteor Society
http://www.amsmeteors.org

Arctic Asteroid!
http://science.msfc.nasa.gov/headlines/y2000/ast01jun_1m.htm

AstroWeb–Astronomy Resources on the World Wide Web
http://www.stsci.edu/science/net-resources.html

Encyclodedia Astronautica
http://www.friends-partners.org/~mwade/spaceflt.htm

NASA Earth Observatory
http://earthobservatory.nasa.gov

NASA Human Spaceflight
http://spaceflight.nasa.gov/station/

NASA: Space Environment Center
http://www.sec.noaa.gov

National Aeronautics and Space Administration
http://www.nasa.gov

Orbital Elements
http://spaceflight.nasa.gov/realdata.elements/index.html

Science News About the Sun-Earth Environments
http://www.spaceweather.com

Sky and Telescope Magazine
http://www.skypub.com/skytel/skytel.shtml

Skyview
http://skyview.gsfc.nasa.gov

Solar Web Guide
http://www.lmsal.com/SXT/html2/list.html

Space News, Games, Entertainment
http://www.spacescience.com

Space Telescope Science Institute
http://www.stsci.edu/

Spacewatch Project, University of Arizona Lunar and Planetary Observatory
http://www.lpl.arizona.edu/spacewatch/index.html

Sunspot Cycle Predictions
http://science.nasa.gov/ssl/PAD/SOLAR/predict.htm

Terra: The EOS Flagship
http://terra.nasa.gov   .

U.S. Naval Observatory
http://aa.usno.navy.mil

Wide Web of Astronomy
http://georgenet.net/astronomy.html

BIOGRAPHIES

African Americans in Science
http://www.princeton.edu/~mcbrown/display/faces.html

American Indian Science and Engineering Society
http://www.aises.org

Biographies of Physicists
http://hermes.astro.washington.edu/scied/physics/physbio.html

4,000 Years of Women in Science
http://www.astr.ua.edu/4000WS/4000WS.html

Galileo and Einstein Home Page
http://galileoandeinstein.physics.virginia.edu

National Academy of Engineering (NAE) Celebration of Women in Engineering
http://www.nae.edu.cwe

Nobel Channel
http://www.nobelchannel.com

Nobel Foundation
http://nobelprize.org/

Nobel Prize Internet Archive
http://www.almaz.com

BIOLOGY

Association for Biology Laboratory Education "Hot" Biology Web Sites
http://www.zoo.toronto.edu/able/hotsites/hotsites/htm

Biolinks.com
http://www.biolinks.com

Biology in the News
http://www.nbii.gov/bionews

Cells Alive
http://www.cellsalive.com

Computer Enhanced Science Education, Whole Frog Project
http://george.lbl.gov/ITG.hm.pg.docs/Whole.Frog/Whole.Frog.html

e-Skeletons Project
http://www.eSkeletons.org

Ecological Society of America
http://www.esa.org/esaLinks.php

Electronic Introduction to Molecular Virology
http://www.uct.ac.za/microbiology/tutorial/virtut1.html

Evolution Website (BBC Education)
http://www.bbc.co.uk/education/darwin/index.shtml

Human Genome Project
http://www.ornl.gov/sci/techresources/Human_Genome/home.shtml

Microbe Zoo—Digital Learning Center for Microbial Ecology
http://commtechlab.msu.edu/sites/dlc-mi/zoo

Microbiology Education Library
http://www.microbelibrary.org

Ongoing Biology
http://www.tilgher.it/ongoing.html

UCMP Exhibit Hall: Evolution Wing
http://www.ucmp.berkeley.edu/history/evolution.html

BOTANY

Agricultural Research Service Science 4 Kids
http://www.ars.usda.gov/is/kids

Ancient Bristlecone Pine
http://www.sonic.net/bristlecone/intro.html

Biotechnology: An Information Resource
http://www.nal.usda.gov/bic

Botanical Society of America
http://www.botany.org

Botany.com
http://www.botany.com

Carnivorous Plants
http://www.sarracenia.com/cp.html

Dr. Fungus
http://www.doctorfungus.org

Introduction to the Plant Kingdom
http://scitec.uwichill.edu.bb/bcs/bl14al.htm

What Is Photosynthesis?
http://photoscience.la.asu.edu/photosyn/education/learn.html

CHEMISTRY

About.com Chemistry
http://chemistry.about.com

Chem Team Tutorial for High School Chemistry
http://dbhs.wvusd.k12.ca.us/webdocs/ChemTeamIndex.html

Chem4Kids
http://www.chem4kids.com

Chemical Education Resource Shelf
http://www.umsl.edu/~chemist/books/

Chemical Heritage Foundation
http://www.chemheritage.org

Chemistry Teaching Resources
http://www.anachem.umu.se/eks/pointers.htm#Curriculum

ChemPen 3D–Classic Organic Reactions
http://home.ici.net/~hfevans/reactions.htm

ChemWeb.com
http://chemweb.com

Classic Chemistry, Compiled by Carmen Giunta
http://web.lemoyne.edu/~giunta

Introduction to Surface Chemistry
http://www.chem.qmw.ac.uk/surfaces/scc

IUPAC Compendium of Chemical Terminology
http://www.chemsoc.org/chembytes/goldbook/index.htm

Molecule of the Month
http://www.bris.ac.uk/Depts/Chemistry/MOTM/motm.htm

On-line Introductory Chemistry
http://www.scidiv.bcc.ctc.edu/wv/101-online.html

Science Is Fun
http://scifun.chem.wisc.edu/scifun.html

COMPUTER SCIENCE AND THE INTERNET

Brain Spin: Technology for Students
http://www.att.com/technology/forstudents/brainspin

Computer Vision Handbook
http://www.cs.hmc.edu/~fleck/computer-vision-handbook

Greatest Engineering Achievements of the Twentieth Century
http://www.greatachievements.org

History of the Web
http://dbhs.wvusd.k12.ca.us/Chem-History/Hist-of-Web.html

Netdictionary
http://www.netdictionary.com

NetLib Repository
http://www.netlib.org

Resource Center for Cyberculture Studies
http://otal.umd.edu/~rccs

Robotics Institute
http://www.ri.cmu.edu

Thinkquest
http://library.thinkquest.org

Virtual Reality
http://www.cms.dmu.ac.uk/~cph/VR/whatisvr.html

EARTH SCIENCES

AgNIC Plant Science Home Page
http://www.unl.edu/agnicpls/agnic.html

ARGO-Observing the Ocean in Real Time
http://www.argo.ucsd.edu

Biota of North America Program
http://www.bonap.org

Botanical Ecological Unit
http://www.fs.fed.us/biology/plants/beu.html

Botanical Electronic News
http://www.ou.edu/cas/botany-micro/ben

Botanical Glossaries
http://155.187.10.12/glossary/glossary.html

Botany
http://www.nmnh.si.edu/departments/botany.html

Botany Online: The Internet Hypertextbook
http://www.biologie.uni-hamburg.de/b-online

Botany.com
http://www.botany.com

Centre for Plant Architecture Informatics
http://www.cpai.uq.edu.au

Delta
http://biodiversity.uno.edu/delta

Digital Tectonic Activity Map
http://denali.gsfc.nasa.gov/dtam

Electronic Sites of Leading Botany, Plant Biology, and Science Journals
http://www.e-journals.org/botany

Food and Agriculture Organization
http://www.fao.org

GardenNet
http://gardennet.com

Global Ice-Core Research
http://id.water.usgs.gov/projects/icecore

Hydrologic Information Center: Current Hydrologic Conditions
http://www.nws.noaa.gov/oh/hic/current

Igneous Rocks Tour
http://seis.natsci.csulb.edu/basicgeo/IGNEOUS_TOUR.html

Index Nominum Genericorum
http://rathbun.si.edu/botany/ing

Integrated Taxonomic Information System
http://www.itis.usda.gov

International Association of Volcanology and Chemistry of Earth's Interior
http://www.iavcei.org

International Organization for Plant Information
http://iopi.csu.edu.au/iopi

International Plant Names Index
http://www.ipni.org

Internet Directory for Botany
http://www.botany.net/IDB

MedBioWorld
http://www.medbioworld.com/bio/journals/plants.html

National Earthquake Information Center
http://wwwneic.cr.usgs.gov

National Oceanic and Atmospheric Administration
http://www.noaa.gov

Natural Perspective
http://www.perspective.com/nature/index.html

Plant Facts
http://plantfacts.ohio-state.edu

Plant Information Systems
http://www.wes.army.mil/el/squa/cdroms.html

Plants Database
http://plants.usda.gov./topics.html

Topozone
http://www.topozone.com

U.S. Geological Survey
http://www.usgs.gov

Virtual Geosciences Professor
http://www.uh.edu/~jbutler/anon/anonfield.html

Virtual Library of Botany
http://www.ou.edu/cas/botany-micro/www-vl

Volcano World
http://volcano.und.nodak.edu

W3 Tropicos
http://mobot.mobot.org/W3T/search/image/imagefr.html

Woods Hole Oceanographic Institution
http://www.whoi.edu

World-Wide Earthquake Locator
http://www.geo.ed.ac.uk/quakes/quakes.html

GENETICS

BioMedNet
http://reviews.bmn.com/?subject=Genetics

DNA from the Beginning
http://www.dnaftb.org/dnaftb

Dolan DNA Learning Center, Cold Spring Harbor Laboratory
http://www.dnalc.org

Genetics Society of America
http://www.genetics-gsa.org

Genome News Network
http://www.genomenewsnetwork.org/index.php

Kimball's Biology Pages
http://biology-pages.info

MedBioWorld
http://www.medbioworld.com

MendelWeb
http://www.mendelweb.org

National Center for Biotechnology Information
http://www.ncbi.nlm.nih.gov

National Public Radio, the DNA Files
http://www.dnafiles.org/home.html

Nature Publishing Group
http://www.nature.com/genetics

U.S. Department of Energy, Genomics Image Gallery
http://www.ornl.gov/sci

U.S. Department of Energy Office of Science, Virtual Library on Genetics
http://www.ornl.gov/TechResources/Human_Genome/genetics.html

University of Massachusetts, DNA Structure
http://molvis.sdsc.edu/dna/index.htm

University of Utah, Genetic Science Learning Center
http://gslc.genetics.utah.edu

HISTORY, SOCIETY, AND CULTURE OF SCIENCE

American Physical Society: A Century of Physics
http://timeline.aps.org/APS/home_HighRes.html

Art of Renaissance Science
http://www.pd.astro.it/ars/arshtml/arstoc.html

Case Studies in Science
http://ublib.buffalo.edu/libraries/projects/cases/case.html

History of the Light Microscope
http://www.utmem.edu/personal/thjones/hist/hist_mic.htm

Important Historical Inventions and Inventors
http://www.lib.lsu.edu/sci/chem/patent/srs136_text.html

Links to Science, Technology, and Society-Related Information Sources
http://www2.ncsu.edu/ncsu/chass/mds/stslinks.html

Science in Our Daily Lives
http://www.lib.virginia.edu/science/events/Sci_Daily_Life.html

INVENTIONS

Invent America!
http://www.inventamerica.com

Invention Convention/National Congress of Inventor Organizations
http://www.inventionconvention.com

Inventors Museum
http://www.inventorsmuseum.com

Inventors Web Site
http://inventors.about.com

Kids Inventor Resources
http://www.InventorEd.org/k-12/becameinv.html

Lemelson-MIT Awards Program Invention Dimension Web Site
http://web.mit.edu/invent/index.html

National Collegiate Inventors and Innovators Alliance
http://www.nciia.org

National Inventors Hall of Fame
http://www.invent.org/book/index.html

Tips for Parents of Young Inventors (Young Inventors Fair Web Site)
http://www.ecsu.k12.mn.us/yif/tips.html

United States Patent and Trademark Office
http://www.uspto.gov

"What It Takes to Be an Inventor": 3M Collaborative Invention Unit
http://mustang.coled.umn.edu/inventing/Inventing.html

MATHEMATICS

Clay Mathematics Institute
http://www.claymath.org

Geometry in Action
http://www.ics.uci.edu/~eppstein/geom.html

Math Archives
http://archives.math.utk.edu

Mathematical Sciences Research Institute
http://www.msri.org/index.html

Metamath Proof Explorer
http://www1.shore.net/~ndm/java/mmexplorer1/mmset.html

MEDICINE AND NUTRITION

American Medical Association
http://www.ama-assn.org

BioMedNet
http://www.biomednet.com

Centers for Disease Control and Prevention
http://www.cdc.gov

Consumer Health Information Service
http://hml.org/CHIS/index.html

Healthlink USA
http://www.healthlinkusa.com

Heart: An Online Exploration
http://sln.fi.edu/biosci/heart.html

Human Anatomy Online
http://www.innerbody.com/htm/body.html

Kids Health
http://www.kidshealth.org

Martindale's Health Science Guide
http://www-sci.lib.uci.edu/HSG/Ref.html

Medicine Net
http://www.medicinenet.com/script/main/hp.asp

Medicine Through Time (BBC)
http://www.bbc.co.uk/education/medicine

Medline Plus
http://medlineplus.gov

Medscape
http://www.medscape.com/px/urlinfo

MEDtropolis Home Page
http://www.medtropolis.com

National Library of Medicine
http://www.nlm.nih.gov

Neuroscience for Kids
http://faculty.washington.edu/chudler/neurok.html

On-line Medical Dictionary
http://www.graylab.ac.uk/omd

WebMD
http://www.webmd.com

World Health Organization
http://www.who.int

MUSEUMS

American Museum of Natural History
http://www.amnh.org

American Museum of Science and Energy, Oak Ridge, Tennessee
http://www.amse.org

California Science Center
http://www.casciencectr.org

Carnegi Museum of Natural History
http://www.CarnegieMuseums.org/cmnh

DNA Learning Center, Cold Springs Harbor
http://vector.cshl.org

Exploratorium
http://www.exploratorium.edu

Field Museum of Natural History in Chicago
http://www.fmnh.org

Franklin Institute Science Museum
http://sln.fi.edu/tfi/welcome.html

Further Explorations
http://www.explorations.org/further_explorations.html

Museum of Science and Industry
http://www.msichicago.org

National Air and Space Museum
http://www.nasm.si.edu

Peabody Museum of Natural History
http://www.peabody.yale.edu

Science Museum of Minnesota, Thinking Fountain
http://www.sci.mus.mn.us

SciTech, the Science and Technology Interactive Center
http://scitech.mus.il.us

Smithsonian Institution
http://www.si.edu

U.S. Space & Rocket Center
http://www.spacefun.com

WebExhibits
http://www.webexhibits.com

PALEONTOLOGY

Coelacanth: The Fish Out of Time
http://www.dinofish.com

Paleonet
http://www.ucmp.berkeley.edu/Paleonet

PaleoQuest
http://paleoquest.cet.edu

So You Want to Be a Paleontologist?
http://www.cisab.indiana.edu/~mrowe/dinosaur-FAQ.html

Willo, the Dinosaur with a Heart
http://www.dinoheart.org

PHYSICS

American Association of Physics Teachers
http://www.aapt.org

American Institute of Physics Center for the History of Physics
http://www.aip.org/history

Elemental Data Index
http://physics.nist.gov/PhysRefData/Elements/cover.html

Exploring Gravity
http://www.curtin.edu.au/curtin/dept/phys-sci/gravity

Gravity Probe B: The Relativity Mission
http://einstein.stanford.edu

Internet Pilot to Physics
http://physicsweb.org/TIPTOP

Particle Adventure
http://www.particleadventure.org

PhysicsEd: Physics Education Resources
http://www-hpcc.astro.washington.edu/scied/physics.html

PhysLINK
http://www.physlink.com

Playground Physics
http://www.aps.org/playground.html

Professor Bubbles' Official Bubble Home Page
http://bubbles.org

Unsolved Mysteries
http://www.pbs.org/wnet/hawking/mysteries/html/myst.html

Visual Quantum Mechanics: Online Interactive Programs
http://phys.educ.ksu.edu/vqm/index.html

SPACE SCIENCE

Air and Space Magazine
http://www.airspacemag.com

Amateur Radio on the International Space Station (ARISS)
http://www.arrl.org/ARISS/

Ames Research Center
http://www.nasa.gov/centers/ames

Ames Research Center: History Office
http://history.arc.nasa.gov

Ames Research Center: Multimedia
http://www.nasa.gov/centers/ames/multimedia

Apollo: A Retrospective Analysis
http://www.hq.nasa.gov/office/pao/History/Apollomon/cover.html

Apollo Program
http://spaceflight.nasa.gov/history/apollo

Apollo Project Archive
http://www.apolloarchive.com

Association of Space Explorers
http://www.space-explorers.org

Astronaut Biographies
http://www.jsc.nasa.gov/Bios

Astronomy Café, The
http://www.astronomycafe.net

Astrophysics Data System
http://adswww.harvard.edu

Cassini-Huygens Mission
http://www.nasa.gov/mission_pages/cassini/main

Chandra X-Ray Observatory Center
http://chandra.harvard.edu

Chronology of Space Exploration, Russian Space Web
http://www.russianspaceweb.com/chronology.html

Coalition for Space Exploration
http://www.spacecoalition.com

Compton Gamma Ray Observatory
http://cossc.gsfc.nasa.gov/cossc

Dawn Mission
http://dawn.jpl.nasa.gov

Deep Impact
http://deepimpact.jpl.nasa.gov

Deep Space Network
http://deepspace.jpl.nasa.gov/dsn

Destination Earth
http://www.earth.nasa.gov

Discovery Program
http://discovery.nasa.gov

Earth Observing System
http://eospso.gsfc.nasa.gov

Earth Observing System Data Gateway
http://delenn.gsfc.nasa.gov/~imswww/pub/imswelcome

European Space Agency (ESA)
http://www.esa.int

Galaxy Evolution Explorer
http://www.galex.caltech.edu

Gamma-ray Large Area Space Telescope (GLAST)
http://glast.gsfc.nasa.gov

Gemini Program
http://www-pao.ksc.nasa.gov/kscpao/history/gemini/gemini.htm

Glenn Research Center
http://www.nasa.gov/centers/glenn

Goddard Space Flight Center (GSFC)
http://www.nasa.gov/centers/goddard

Great Images in NASA
http://grin.hq.nasa.gov

History Timelines
http://www.hq.nasa.gov/office/pao/History/timeline.html

Human Space Flight
http://spaceflight.nasa.gov/home

International Space Station
http://www.nasa.gov/mission_pages/station/main

Jet Propulsion Laboratory (JPL)
http://www.jpl.nasa.gov/

Johnson Space Center (JSC)
http://www.nasa.gov/centers/johnson

Kennedy Space Center
http://www.nasa.gov/centers/kennedy

Konstantin E. Tsiolkovsky State Museum of the History of Cosmonautics
http://www.informatics.org/museum

Langley Research Center (LRC)
http://www.nasa.gov/centers/langley

Lunar and Planetary Science
http://nssdc.gsfc.nasa.gov/planetary/planetary_home.html

Lunar Exploration Timeline
http://nssdc.gsfc.nasa.gov/planetary/lunar/lunartimeline.html

Lyndon B. Johnson Space Center
http://www.nasa.gov/centers/johnson

Manned Spacecraft Gallery
http://www.lycoming.edu/astr-phy/fisherpdg1

Mars Exploration Program
http://mars.jpl.nasa.gov/

Marshall Space Flight Center
http://www.nasa.gov/centers/marshall

Mercury, Project
http://www-pao.ksc.nasa.gov/kscpao/history/mercury/mercury.htm

Mir Space Station, Russian Space Web
http://www.russianspaceweb.com/mir.html

NASA
http://www.nasa.gov

NASA: History Office
http://history.nasa.gov/

NASA: Image Exchange (NIX)
http://nix.nasa.gov

NASA: Science Mission Directorate
http://science.hq.nasa.gov/

NASA: Space Science
http://spacescience.nasa.gov

New Millennium Program
http://nmp.jpl.nasa.gov

Origins of the Universe
http://origins.jpl.nasa.gov

Planetary Society, The
http://planetary.org

PlanetScapes
http://planetscapes.com

Satellite Tracking in Real Time
http://science.nasa.gov/realtime

Skylab Program
http://www-pao.ksc.nasa.gov/kscpao/history/skylab/skylab.htm

Solar System Exploration
http://sse.jpl.nasa.gov

Spaceline.org History of Rocketry
http://www.spaceline.org/rockethistory.html

Spitzer Space Telescope
http://www.spitzer.caltech.edu/spitzer

Stardust Project
http://stardust.jpl.nasa.gov

Stennis Space Center, John C.
http://www.nasa.gov/centers/stennis

TDRSS Satellite System
http://msp.gsfc.nasa.gov/tdrss/tdrsshome.html

Ulysses
http://ulysses.jpl.nasa.gov

Virtual Space Museum (of Cosmonautics)
http://vsm.host.ru/emain.htm

Wilkinson Microwave Anisotropy Probe
http://map.gsfc.nasa.gov

WEATHER, CLIMATE, AND NATURAL DISASTERS

Environmental Protection Agency, Global Warming Site
http://www.epa.gov/globalwarming

Global Volcanism Program
http://www.volcano.si.edu/gvp/index.htm

Hurricane and Storm Tracking
http://hurricane.terrapin.com

Lightning Imaging Sensor Data
http://thunder.msfc.nasa.gov/data/lisbrowse.html

National Drought Mitigation Center
http://enso.unl.edu/ndmc

National Snow and Ice Data Center
http://www-nsidc.colorado.edu/NSIDC

National Weather Service
http://www.nws.noaa.gov

Nova Online: Flood!
http://www.pbs.org/wgbh/nova/flood

Tornado Project Online
http://www.tornatoproject.com

Tsunami!
http://www.geophys.washington.edu/tsunami/intro.html

ZOOLOGY

AmphibiaWeb
http://elib.cs.berkeley.edu/aw

Animal Diversity Web
http://animaldiversity.ummz.umich.edu

Endangered Species Update
http://www.umich.edu/~esupdate

Extinction Files
http://www.bbc.co.uk/education/darwin/exfiles/index.htm

FishBase
http://www.fishbase.org

Ichthyology Web Resources
http://www.biology.ualberta.ca/jackson.hp/IWR/index.php

Living Links
http://www.emory.edu/LIVING_LINKS

National Marine Mammal Laboratory's Education Web Site
http://nmml.afsc.noaa.gov/education

NetVet and the Electronic Zoo
http://netvet.wustl.edu

North American Bird Conservation Initiative
http://www.bsc-eoc.org/nabci.html

Northern Prairie Wildlife Research Center
http://www.npwrc.usgs.gov

Satellite Tracking of Threatened Species—NASA
http://sdcd.gsfc.nasa.gov/ISTO/satellite_tracking

Virtual Creatures
http://k-2.stanford.edu/creatures

Welcome to Coral Forest
http://www.blacktop.com/coralforest

World Wildlife Fund
http://www.wwf.org

—Elizabeth Schafer, updated by Christina J. Moose

SCIENCE
AND
SCIENTISTS

Category Index

CHEMISTRY

COMPUTER SCIENCE

EARTH SCIENCE

Antisepsis, 23
Aspirin, 31
Blood Circulation, 89
Blood Groups, 95
Blue Baby Surgery, 99
Contagion, 187
Diphtheria Vaccine, 236
Galen's Medicine, 345
Germ Theory, 382
Greek Medicine, 414
Hand Transplantation, 425
Heart Transplantation, 429
Human Anatomy, 462
Human Genome, 470
Human Immunodeficiency
    Virus, 474
Hybridomas, 477
Immunology, 490
In Vitro Fertilization, 496
Insulin, 512
Manic Depression, 604
Microscopic Life, 637
Mitosis, 645
Neurons, 664
Oncogenes, 687
Ova Transfer, 704
Penicillin, 729
Polio Vaccine: Sabin, 773
Polio Vaccine: Salk, 777
Psychoanalysis, 796
Pulmonary Circulation, 801
REM Sleep, 863
Schick Test, 887
Smallpox Vaccination, 914
Stem Cells, 950
Streptomycin, 965
Viruses, 1002
Vitamin C, 1005
Vitamin D, 1009
Yellow Fever Vaccine, 1052

METEOROLOGY
Atmospheric Circulation, 35
Atmospheric Pressure, 39

Chaotic Systems, 160
Weather Fronts, 1027

METHODS AND INSTRUMENTS
Celsius Temperature Scale, 146
Fahrenheit Temperature Scale,
    318
Herschel's Telescope, 443
Hubble Space Telescope, 455
Kelvin Temperature Scale, 547
Longitude, 590
Periodic Table of Elements, 733
Radiometric Dating, 850
Recombinant DNA
    Technology, 854
Schick Test, 887
Scientific Method: Aristotle,
    894
Scientific Method: Bacon, 899
Scientific Method: Early
    Empiricism, 905
Spectroscopy, 927
Very Long Baseline
    Interferometry, 998
X-Ray Crystallography, 1043
X-Ray Fluorescence, 1047

PALEONTOLOGY
Amino Acids, 14
Australopithecus, 55
Cro-Magnon Man, 209
Evolution, 300
Fossils, 329
Gran Dolina Boy, 393
Human Evolution, 465
Lamarckian Evolution, 560
Langebaan Footprints, 566
Lascaux Cave Paintings, 568
Lucy, 595
Microfossils, 633
Neanderthals, 658
Peking Man, 720
Qafzeh Hominids, 809
Zinjanthropus, 1055

Personages Index

Subject Index

Atomic bomb, 683, 770
Atomic clocks, 656, 1000
Atomic mass, 51
Atomic nucleus, 42-47, 542, 677;
  binding force, 831. *See also*
  Nuclear fission
Atomic number, 272
Atomic pile, 681-682
Atomic structure, 10, 47-50, 433,
  541, 826; exclusion principle,
  306-310
Atomic theory, Greek, 43
Atomic theory of matter (Dalton),
  50-55
Atomic weights, 734
Aubrey, John, 955
Aurignacian period, 213, 571
Auroras, 919
Ausman, Neal E., Jr., 351
Australopithecus, 1056
Australopithecus afarensis, 599
Australopithecus africanus, 55-59,
  596, 723
Australopithecus boisei, 596,
  1055-1059
Australopithecus robustus, 596
Avebury, 955
Averroës, 626
Avery, Oswald T., 252
Axiom of choice (Levi), 59-64
Axiomatic method, 617
Axioms of Euclid, 297
Axioms of motion (d'Alembert),
  213-217

Baade, Walter, 310, 671, 839, 948
Bacon, Francis, 600, 604, 900, 903
Bacteria; discovery, 640; disease-
  causing, 191, 237, 385, 887,
  1002; disease-fighting, 965;
  genetic engineering, 855

Baire, René, 62
Ballard, Robert D., 485, 487
Ballistics, 64-69, 324
Balloons, 203, 206
Baltimore, David, 688
Banting, Frederick Grant, 451,
  513
Bar Kokhba, 220
Baran, Paul, 530
Bardeen, John, 972, 976
Bardeen, William, 818
Bardon, L., 658
Barghoorn, Elso, 14, 634
Barkla, Charles Glover, 1047
Barnard, Christiaan, 429
Barnett, Miles A. F., 535
Barometers, 39-42
Barrow, Isaac, 126, 406
Barthelmy, Scott, 359
BASIC computer language, 740
Bateson, William, 632
Bauhin, Gaspard, 81
Bayer Company, 33
Bayliss, William Maddock, 451
Beagle expedition, 301
Becquerel, Antoine-Henri, 1048
Bednorz, Johannes Georg, 975
Bees, 509-512
Behring, Emil von, 236, 492, 887
Bei, Weng Zhong, 722
Beijerinck, Martinus, 1002
Bell, Jocelyn, 673, 805, 846
Bell curve, 69-74
Bell Laboratories, 202, 574, 836
Benedetti, Giovanni Battista, 65
Beppo-Sax (telescope), 361
Berg, Paul, 248, 854
Berger, Lee, 566
Bergeron, Tor, 1028
Bergstralh, Jay, 351
Bernard, Claude, 385